Exposing the Great Climate Change Lie

Lynne Balzer

Faraday Science Institute

FARADAY SCIENCE INSTITUTE
"A BETTER WORLD THROUGH SCIENCE AND TECHNOLOGY"

Dedication

Fred Singer (1924 – 2020), a scientist who made key advances in rocketry and atmospheric research, was one of the most distinguished atmospheric physicists in the United States. He spoke out about the invalid hypothesis of global warming.

Dr. Singer was a Princeton-trained physicist who held a number of prestigious academic posts throughout his career. He was a longtime professor of environmental sciences at the University of Virginia.

westmiller.com

He conducted some of the first experiments with high-altitude rockets and satellites, measuring cosmic rays and other components in the upper atmosphere. Working with the National Oceanic and Atmospheric Administration (NOAA), Dr. Singer participated in early efforts to use satellites in weather forecasting.

Dr. Singer received a US Dept. of Commerce Gold Medal award for his leadership. With Dennis T. Avery, he wrote *Unstoppable Global Warming: Every 1500 Years*, published in 2008, and many other books. He was also founder and chairman of the Science and Environmental Policy Project.

Dr. Singer will be sorely missed for his outstanding contributions to the field of atmospheric physics and for his insightful observations about the climate.

Foreward

We are living in a very dangerous time in world history. The climate change discussion, which has been going on for nearly fifty years, centers around energy, because energy is what the world's people need to survive and thrive. Governments have realized that energy is power. And therefore, whoever controls energy controls the people.

An evil force has been taking steps to curtail the amount of energy available for use by the world's people. Those in control have accomplished this by limiting the oil and coal available. In the United States, the Biden regime put an end to oil and natural gas drilling by either canceling or over-regulating drilling leases. This policy was started on Joe Biden's first day in office, when he signed an executive order canceling the Keystone pipeline. The progressive Democrats found a way to institute the "Green New Deal" without authorization from Congress: make fossil fuel energy so expensive that "green" energy seems more appealing.

By June of 2022 the price of oil had risen to about $122 per barrel. But there is no shortage of oil, even though the cartels running the world would like you to believe that there is. There is an estimated 14 trillion barrels in the ground waiting to be pumped. And that is just what has been discovered so far. Yet, all the petroleum used by the world's people since its discovery in 1859 amounts to about one trillion barrels of oil.

People who work in the oil fields have told us that it costs only about five or ten cents to pump one barrel of petroleum. Whether that figure is accurate or not, it's obvious that the real cost of petroleum is much, much lower than the amount the oil companies are charging for it.

Petroleum is used not just for transportation - but also for fertilizer, without which our world's food supply will take a tremendous hit. Other products made from petroleum are plastics and rubbers, soaps and detergents, solvents, drugs, pesticides, explosives, synthetic fibers, paints, epoxy resins, and flooring and insulating materials. Interfering with oil production is causing prices to skyrocket and wrecking economies. This is happening in other parts of the world as well. It is not an accident.

The Biden regime has made it very clear that their goal is to limit the availability of "fossil fuels" to drive energy prices so high that people will be ready to accept alternatives, the so-called "renewables", which, incidentally, are not renewable at all. (This book explains why.)

Meanwhile Germany and some other European countries are scrambling to recover their traditional energy sources such as coal power plants and firewood so that their people won't freeze to death this winter. They've finally realized that the energy alternatives they've been installing - wind, solar and biofuels – presented in glowing terms decades ago, are not doing the job they were set up to do.

During the past year the wind didn't blow as expected across Europe. So the expensive wind turbines they'd built didn't produce the energy they hoped for. People were forced to use reserves of natural gas normally saved for winter. This set up a cascading series of shortfalls leading to dramatically higher energy prices and the prospect of shortages and blackouts this coming winter.

These people were led down the primrose path to disaster by articles such as the one in Bloomberg, which stated, "Germany plans to rapidly accelerate the expansion of wind and solar power...by 2035." Just three years ago Forbes stated that renewable energy is now the "cheapest option". And in June of 2021 their headline read, "Building New Renewables is Cheaper Than Burning Fossil Fuels." This couldn't be further from the truth. The London Financial Times reported on July 8, 2022, that EU electricity prices had quadrupled over the last year, reaching the highest level on record.

The US should take heed from the experience of the people of Europe, Sri Lanka, Ghana and South Africa, who are the "canaries in a coal mine." Otherwise the same thing will ultimately occur here. It is critical that Americans wake up to the truth to prevent the economic destruction of the US.

This book will give you a basic working knowledge of the scientific facts about climate to enable you to recognize and effectively deal with climate change propaganda. I hope you will not just take my word for this. Please do some research to verify the truths this book presents.

Contents

Chapter One

The Science

> Key to Science: "If it disagrees with experiment, it is wrong. In that simple statement is the key to science. It doesn't make any difference how beautiful your guess is. It doesn't make any difference how smart you are, who made the guess, or what his name is. If it disagrees with the experiment, it's wrong. That's all there is to it." - Professor Richard Feynman, Nobel Laureate

Everyone acknowledges that the earth is experiencing climate change. The climate has been changing since the beginning of Earth's history 4.5 billion years ago. Since the beginning of geologic history there have been approximately six great ice ages. We're now actually between two parts of the Pleistocene Ice Age.

The two questions we need to ask about this are:

(1) "How much climate change has occurred since the beginning of the Industrial Revolution?" (around the year 1800) and

(2) "How much of this change is attributable to humans and how much of it to natural causes?"

This book presents the facts you need to help you explore these important questions. Despite what you may have heard from the mainstream media, the science has *not* been settled, despite the *trillions* of dollars that have now been spent on researching and dealing with this matter. Not only have taxpayers been saddled with an almost unbelievable expense, but — even worse — we are now witnessing the loss of basic freedoms of speech and movement being imposed upon us in the name of climate change.

Each year the world's richest people, members of the World Economic Forum (WEF), hold a conference in Davos, Switzerland. There they discuss their latest plans, such as requiring every resident of this planet to record his/her "carbon footprint". This is the amount of carbon dioxide a person generates in daily activities: miles traveled, food eaten, items purchased, etc. They are now threatening the world's food supply by declaring nitrates, used in fertilizers, bad.

Untangling Truth from Falsehood

Many people feel incompetent to investigate these questions on their own, relying instead on media they trust. The videos and articles presented by the global alarmists are scarce on evidence or citations. They tell us to rely on “experts”. They rarely attempt any proof to back up their statements. They know that it’s easier for the public to swallow these pronouncements hook, line and sinker than to check them out using critical thinking, which asks the questions “why” and “how.”

A recent study revealed that most people feel uncomfortable with science and are thus afraid to question the authorities. This is because too many of us were taught to view science as a long list of difficult equations and definitions rather than an exciting exploration of the world around us.

Climate is average weather over a period of time, and weather has a great influence on our daily activities. So everyone should understand basic facts about weather. Yet recently 50% of a group of Yale University students failed a test of basic climate knowledge. It’s no wonder that most people feel inadequate to question what we’re being told about changes in the climate and what we can do — if anything — about it.

First we need to understand how science works. Science is more than just knowledge about our natural environment. It’s also a way of thinking and approaching problems. If the rules for scientific investigation are not followed, the science is flawed, and the conclusions reached are probably erroneous. Science isn’t omnipotent; it has its limitations. And scientists aren’t little gods. They are human beings with the same tendency to be corrupted as everyone else. A true scientist is humble enough to admit s/he was wrong about something. The smugness we are seeing today among global warming alarmists is antithetical to science. "It's almost like a religious creed. Science isn't faith. It's rigorous skepticism." [1]In the COVID-19 dilemma, a similar situation, Senator Ron Johnson correctly observed, "Research grants have corrupted medical science." This is also true of climate science. And now those in power are censoring anyone who disagrees with the official position.

The four main tenets of the human-caused [anthropogenic] global warming hypothesis are:

1) “Carbon dioxide [CO2] is the principal greenhouse gas.”

This is a lie. Water vapor, making up about 3% of the atmosphere, is the most powerful “greenhouse gas”. At 30,000 parts per million (ppm), water vapor has 70-100 times as much effect on the warming of the atmosphere as carbon dioxide, which, at 415 ppm, makes up only about 0.04% of the atmosphere.

2) “Carbon dioxide controls global temperature, which has been steadily increasing.”

To find out if this is true, we first need to examine previous weather reports to learn how much the average global temperature has increased in the last 100 years. We also need to know how much of this has been natural change and how much of it was caused by humans.

3) "Industrialized nations especially must atone, or pay the price, for this warming."

This makes it very clear that climate change is a political — not a scientific — issue.

4) "Man can control the climate. We can fix this problem by emitting less CO_2."

If it's true that mankind's addition of CO_2 to the atmosphere has caused climate change above and beyond natural causes, then theoretically we could stop it. But has it ever been proved that carbon dioxide is the culprit?

Climate Alarmism and the Environmentalist Cause.

About 80% of a group of people recently surveyed said that they care deeply about the environment. Proponents of Climatism, the belief in anthropogenic (human caused) global warming (AGW), who say that without drastic action the world will end in ten years, have taken the moral high ground by attaching this issue to environmentalism. This has actually become a religion for many people.

Fortunately the Internet has made it easier than ever before to research this claim to verify its accuracy. Fifty years ago one would have had to comb through hundreds of books and periodicals in a huge library to get the same information that can now be obtained in just a few hours online.

But powerful forces are already closing some doors to the truth by censoring material that doesn't follow the official narrative — just as they've censored doctors warning us about COVID-19 Mrna "vaccines". So, to find the truth, we have no time to lose in checking out the facts about this vital matter that concerns us all. Hopefully you will check claims of impending doom caused by human-generated carbon dioxide and — almost as important — the probability that the proposed alternatives will work.

If you lack the confidence to make an informed decision about this important issue because you feel that you lack scientific knowledge or qualifications, this book will provide you with a clear and honest presentation of the facts. You don't need a PhD in climate science to grasp the issue when it's presented in concrete, down-to-earth terms.

Science or Pseudoscience?

The following have been blamed on climate change: more heat waves but also more cold spells, more droughts but also more floods, less snow but also more snow, more mental illness and suicides, more alcoholism, more crime, more bugs, rats and bats, the civil war in Syria, more earthquakes and tsunamis.

A scientific hypothesis is an idea that guesses the cause of some phenomenon of nature that has been observed. Karl Popper (1902–1994) once stated, "If a theory can be falsified, it is science. Otherwise it is pseudoscience [false science]." Since whatever happens with the weather is now blamed on "climate change," there is no way to disprove this hypothesis. If there is a heat wave, we are told that it is caused by climate change. If there is a severe cold snap, that is also blamed on human-generated climate change. Thus it is impossible to prove that human-caused (anthropogenic) climate change is not happening. Therefore, the hypothesis is unscientific.

The burden of proof that human emission of CO_2 is causing climate change rests on those who have been calling this hypothesis a fact for the past forty years, since they made this claim in the first place. To this date absolutely no evidence has been provided that the trace gas carbon dioxide is a significant driver of climate. A $10,000 reward was offered to anyone who can provide this evidence. But no one has responded to this challenge.

It's true that the percentage of carbon dioxide, making up only about *four hundredths of one percent* (0.0415%) of the atmosphere, has been rising at a steady pace since the mid 20th century. The reasons for this increase in carbon dioxide, only 4% of which was caused by mankind's activities, are explained in Chapter 3.

It is a fact that during the 1960s and 1970s, while the percentage of carbon dioxide was increasing, the earth was getting cooler. So who could then make the assumption that increasing the amount of carbon dioxide would cause temperatures to rise? Empirical (evidence-based) science records data from observations, using scientific instruments. Piers Corbin, a British meteorologist noted for his long-range weather forecasts, stated, "There are no scientists in the world who can produce such observational data from the real world...There is only effect the other way, namely that the ocean temperatures control average CO_2 levels due to basic laws of physics about the solubility of CO_2 gas in [sea] water." [2](See Chapter 3)

Computer Models

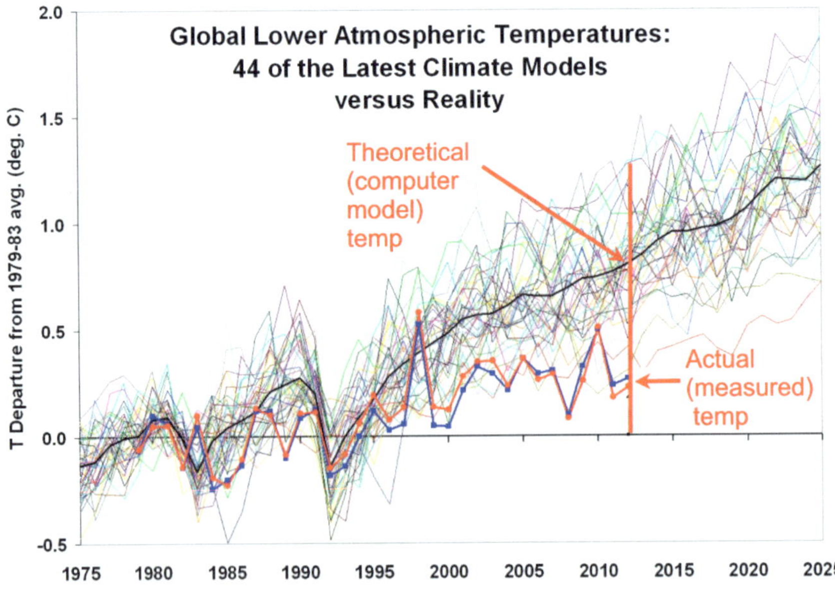

Computer model predictions are inaccurate..

Predictions of the amount of future warming on which the United Nations International Panel on Climate Change (IPCC) bases its claims hinge on data obtained by approximately 100 computer models. The problem with models is that if you don't correctly program into the computer everything, including the tiniest detail, involved with weather and climate, you cannot expect a valid result. Weather patterns are very complicated, and climatologists admit that many of these factors are not well understood. This is why predictions of future temperatures made by computer models have been so far off base. The warming they've predicted has been about three to four times as much as has actually occurred in the last forty years since they started making these predictions. To reduce the amount of error, the people putting information into these models have periodically tweaked, or tuned them to conform to past records. In spite of this, they are still off by a great deal, as you can see on this graph.

Dr. John Christy, University of Alabama professor and Alabama State Climatologist, discovered that when the model warms 1K, it allows 1.4 watts of heat to escape. But when the temperature rises 1K in the real world, 2.6 watts of energy are released. Thus, the model derives an unrealistic amount of warming. Also, computer models are programmed to assume that carbon dioxide is the largest climate driver. In science we measure and record. Computer models do not measure anything. The results obtained by these models depend on what was programmed into them.

The University of Alabama maintains the most up-to-date, accurate database of temperature information collected by satellites. This database reports 0.136 degrees C of warming per decade, which amounts to a 1.4K temperature increase per century.

True empirical science is based on observations and measurements, using our five senses and instruments, of what is happening in the real world (not a computer model). In order to study weather and climate, temperature, rainfall, wind speed and direction, air pressure and sea ice extent must be carefully measured, monitored, and recorded over a significant period of time.

The pieces of information gained in these measurements and tests are called data (a plural word). Data are carefully analyzed and used to draw conclusions and make predictions. This procedure is known as the scientific method. If the results or predictions of a hypothesis (initial idea) turn out to be incorrect, then it follows that either the hypothesis or the way of testing it is wrong. If the results conflict with the hypothesis, then it must be scrapped or changed. There's nothing wrong with changing your mind about an issue when new data come to light.

After a report of a scientific study is carefully written by the researchers, they must submit it, along with their data, to a scientific journal for publication. This important process gives other scientists in that field (called peers) the opportunity to read and evaluate the study's conclusions and methods. Other scientists repeat the investigation, which should yield a similar result. This essential step of the scientific method is called peer review.

Because observations and experiments must be repeated over and over again, scientific investigation is difficult and time consuming. It could take years to complete a study and more years for other scientists working in the same field to verify the results. Only then can a hypothesis be considered a "theory". Established theories are questioned and sometimes overturned. It is the very nature of science to be skeptical. While computer models could be helpful at times, they are no substitute for empirical science.

In 2012 forty nine former NASA scientists, including seven astronauts, sent a letter to NASA expressing their dismay that this agency — specifically the Goddard Institute for Space Studies (GISS) — was drawing conclusions based on scientifically flawed climate models that had failed to make accurate predictions and were ignoring proper scientific methods of investigation.[3] Some scientists have expressed their concern that when the truth about this deception does come out in the open, many people who once believed in the integrity of science will no longer trust science for answers. This is already happening, since it is now obvious that scientists lied to us about the origins of COVID-19, and the treatments for this disease were hidden from the public through censorship, causing the deaths of millions of people.

For decades scientists have been handsomely paid with government grants to perpetuate a human-generated crisis. If they stop doing this, their grants will be discontinued. In other words, they'll be out of a job. No wonder they've come to these startling and incorrect conclusions.

Climate data have been changed.

As unbelievable as it may seem, agencies in the United States, Australia, New Zealand and other nations have actually *changed their own temperature records*. Many climate scientists have been surprised and shocked to see later editions of graphs published by the National Oceanic and Atmospheric Administration (NOAA) that had been changed from their previous editions.

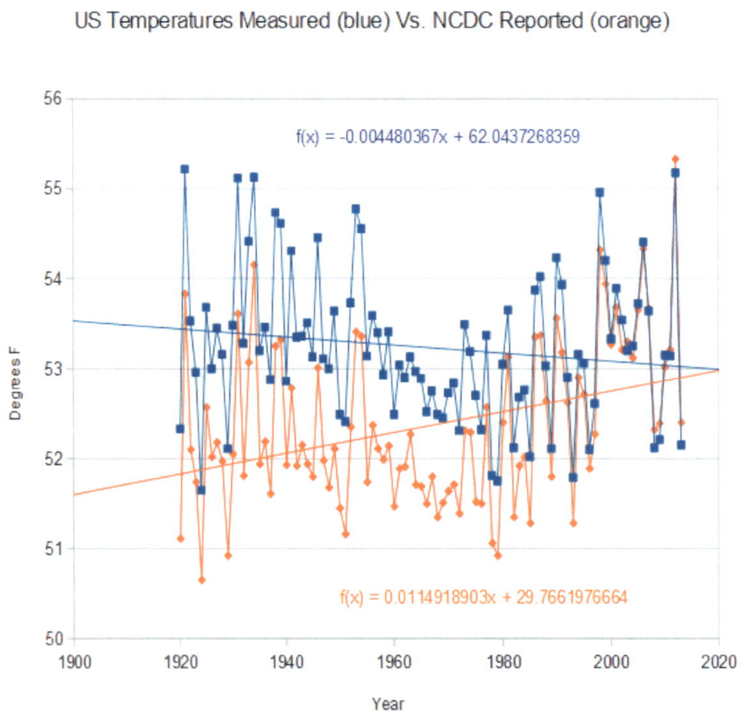

The National Oceanic and Atmospheric Administration (NOAA) and NASA actually lowered recorded temperatures from the early 20th century and raised more recent temperatures, making it appear that temperatures have risen. Just as in the novel *1984*, they've actually changed history. This is not a conspiracy theory but an easily-provable fact, determined by comparing previously published graphs with the newer editions of the same graphs published later by these agencies after they had changed the data.

The blue line in this graph shows the actual measured temperatures downloaded by scientists from official government websites. The orange line representing data that was changed was shown on the graph ultimately released to the public. As you can see, the temperatures from the early part of the 1900's were lowered from what they actually were. The truth is that in the 1920s and 1930s temperatures were similar to today's and even warmer.

In the year 1911 there was a terrible heat wave that killed tens of thousands of people in Europe and the United States. This is verified by looking at newspaper articles from 1911. In France alone 41,000 people died.[4] In Nashua, New Hampshire, the temperature reached 106° F. Yet strangely, NASA's graph indicates that 1911 was a cold year.

Tony Heller (RealClimateScience.com), who has carefully researched this matter, concluded that NOAA's "adjustments" actually represent the entire 'warming trend'".[5] This is worse than lying with statistics. It is entirely made up. These agencies don't deny that they've cooked the data. With a flurry of double talk they admit that they've made "adjustments" to it. "In the year 2000 NASA charts showed about 0.5°C. of warming prior to 2000. Now, in 2023, the same graphs show nearly 1.5°C warming prior to the year 2000".[6] This demonstrates how they have "adjusted" the temperature data.The same methods of fudging temperature data under the direction of the Climate Research Unit (CRU) in Britain were also used in Australia and New Zealand. This action goes against every principle of science. It is blatant fraud being committed right before our eyes.

If the "US Heat Waves" graph shown below had been superimposed on the climate record starting with the year 1900, it would have come out looking like the correct graph below it. Notice how much more severe the heat waves were in the 1930s and 1940s.

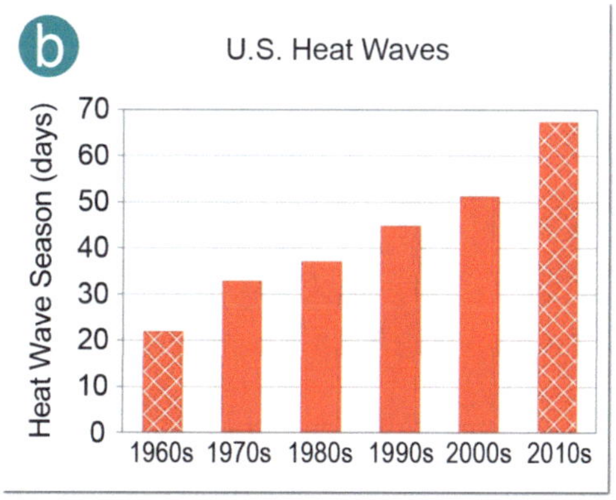

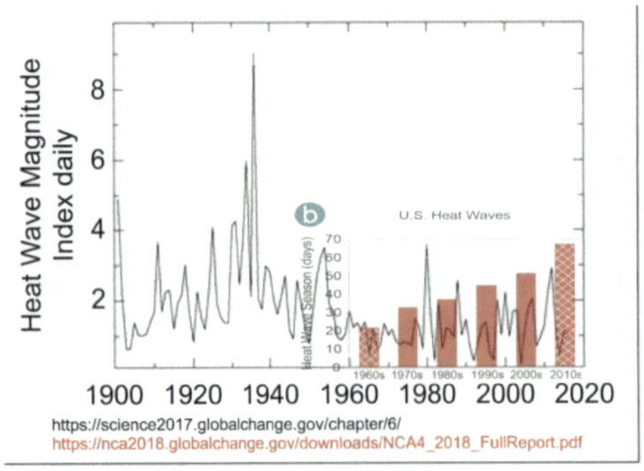

Why did the graph they released in 2018 show the 1960s as their starting point? Because the decade of the 1960s was the coolest of the past century. "The 2018 report released to the public cherry-picked the only start date in the graph which would create the appearance of a warming trend."[7] They simply acted as if the records previous to the 1960s didn't exist. This is one way of lying with statistics and is fraud of the highest order.

Here's a graph of temperature records for Reykjavic, Iceland, from 1900 on. The red arrow shows the time at which a graph showing the temperatures began. This totally eliminated the warmer period occurring in the 1930s and 1940s. Climate is cyclical, and anyone who studies this for any length of time can't miss that fact.

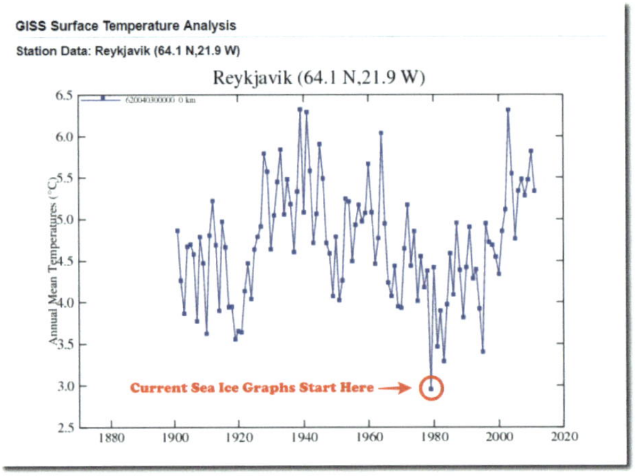

The two graphs below depict measured temperatures from 1996 to 2016. The 2016 edition shows the temperatures as they were measured. The 2020 edition of the same graph reflects temperatures that have been modified or "adjusted". These changes they made are not small changes. They are huge. The modified graph shows a steep rate of warming.

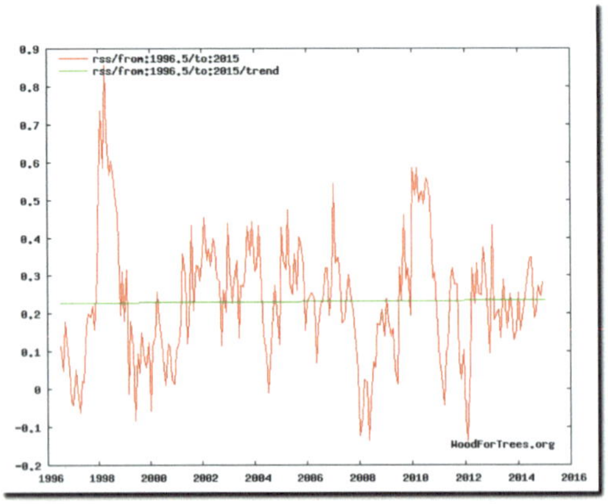

Temperature Graph - 2016 Edition

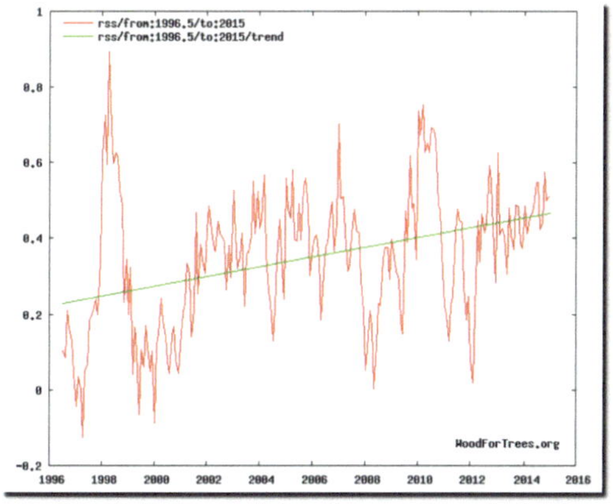

2020 Edition of the Same Graph (After Adjustments)

Thomas Karl, of the National Oceanic and Atmospheric Administration's National Climate Data Center in Asheville, NC, observed, "Analysis of warming since 1881 shows most of the increase in global temperature happened before 1919 — before the more recent sharp rise in the amount of carbon dioxide in the atmosphere... While global climate warmed overall since 1881, it actually cooled from 1921 to 1979."

In 2009 Dr. Alan Carlin, an employee of the Environmental Protection Agency (EPA), was asked to review the Draft Technical Support Document for Endangerment Analysis for Greenhouse Gas Emissions under the Clean Air Act. In an 85-page paper he made nineteen recommendations for revisions to that document. He and Dr. John Davidson criticized the EPA for accepting the Intergovernmental Panel on Climate Change's (IPCC's) mistaken conclusions and failing to do their own scientific research. Some of their conclusions were:

> "Global temperatures have declined extending the current downtrend to years. At the same time atmospheric CO_2 levels have continued to increase and CO_2 emissions have accelerated...The consensus on past, present and future Atlantic hurricane behavior has changed...Now the consensus is much more neutral, arguing that future Atlantic tropical cyclones will be little different than those of the past...The idea that warming temperatures will cause Greenland to rapidly shed its ice has been greatly diminished by new results indicating little evidence for the operation of such processes...Given the downward trend in temperatures since 1998 (which some think will continue until at least 2030), there is no particular reason to rush into decisions based upon a scientific hypothesis that does not appear to explain most of the available data."[8]

However, a senior EPA official, Al McGartland, quickly put the lid on this paper, admonishing its writers in an email, "The administrator and the [Obama] Administration have decided to move forward [with a cap-and-trade bill] ...and your comments do not help the legal or policy case for this decision."

Three Equations Debunking the AGW Narrative

Recently Christopher Monckton, a former advisor to British Prime Minister Margaret Thatcher and renowned mathematician published in peer-reviewed climate journals, presented three equations that he believes will destroy the climate change alarmist narrative.[9]

The first equation corrects the equation climate scientists have used to calculate the amount of warming we'd have if the percentage of CO_2 (0.04%) doubled. Moncton explains that the equation climatologists have been using is incorrect, because it fails to start with the baseline: Earth's temperature *without* greenhouse gases. Without the greenhouse effect, Earth's average temperature would only be about 260K (-13 degrees C). This number must be in the equation as a baseline signal. The corrected equation is:

260K (original temperature) + 29K (total greenhouse effect) ÷ 260K + 9K (direct greenhouse warming) = 1.1

This number turns out to be only about one third of the result obtained when the baseline signal is not used in the equation. In other words, this would amount to much less warming.

The second equation centers on the fact that as the percentage of "renewable" energy feeding an electrical grid increases, a point is reached at which the energy input at various times (mainly in the afternoon) exceeds the current demand, and the energy is thus wasted. This is known as the "duck curve", discussed in Chapter 8. To meet the 50% goal governments have mandated, every wind or solar power plant would have to be overbuilt to a ridiculous extent. In this case renewable energy becomes prohibitively expensive.

Douglas Pollock, a Chilean engineer who discovered this phenomenon, was blocked from warning Latin American policymakers about this serious problem. Yet this problem has already surfaced in California and in seven countries. [10]Douglas Pollock's equation is:

E (excess power) = N (nameplate capacity) – D (hourly demand)

The excess power is wasted, because hardly any of it can be saved in batteries. This is because battery technology has failed to keep up with the requirements for such a system. So these power companies will never be able to reach 50%. It makes no sense to mandate 50% unless the policy makers setting these requirements are hell bent on destroying economies.

The third equation involves the amount of warming that would be prevented by 2050 if the entire world produced no net CO_2 emissions from now until that time. The equation is:

Warming prevented = 0.45 units of influence abated x 1.8 TCR (transient doubled CO_2 response) ÷ 3.93 ERF (effective radiative forcing) units = 0.2K.

Therefore, if somehow all human-caused CO_2 emissions were to immediately cease, by the year 2050 just 0.2 of a degree of warming would be averted. This would cost many trillions of dollars, even if possible, and no one would even notice 0.2 degrees. Christopher Monckton pointed out that "The UK national grid has estimated $3.6 trillion as the cost of re-engineering the grid to meet the net-zero target."

Summary

The scientific basis for the climate change hypothesis has failed, because

- Computer models are no substitute for real-world empirical measurements. The figures obtained by these models have consistently predicted a much greater degree of warming than has been actually observed during the nearly forty-year period in which they've been used. As they say, "Garbage in, garbage out."
- Agencies such as NASA and NOAA have changed historical temperature records and dishonestly manipulated graphs. Reports by scientists differing from the official conclusions have been ignored.
- Incorrect equations have been used to reach some of these conclusions.
- Climate change has been blamed for every extreme weather event, even though extreme weather events such as hurricanes and tornadoes, have occurred throughout history. Thus, the Anthropomorphic Global Warming hypothesis cannot be disproved and should be scrapped.

It has never been definitively proved that CO_2, a trace gas making up only about four hundredths of a percent of the atmosphere, is a direct driver of the slight warming that has occurred since the year 1800. This will be discussed further in Chapter 3.

---

1. Ben Bikman, PhD
2. WeatherAction.com
3. Cfact, "Astronauts and scientists send letter to NASA: Stop global warming advocacy," www.cfact.org, April 10, 2012
4. https://climatechangedispatch.com/france-1911-heatwave-41000-deaths/
5. Tony Heller, "Climate Scientists Rewriting The Past," www.realclimate-science.com/climate-scientists-rewriting-the-past/
6. Ibid.
7. Ibid.
8. "EPA Dept. Director Kills Report finding Global Warming is an invalid hypothesis", July 7, 2009, climetechsolutions.com
9. Christopher Monckton, "Three Damning Equations To Defeat the Global Warming Zealots," American Thinker, May 12, 2023
10. Germany, Spain, Ireland, the United Kingdom, Italy, Australia and Chile

Chapter Two

Climate Alarmism History

> "We (UN-IPCC) redistribute de facto the world's wealth by climate policy. One has to free oneself from the illusion that international climate policy is environmental policy. This has almost nothing to do with environmental policy anymore." - Ottmar Endenhofer, IPCC Cochair of Working Group 3

Climate Alarmism Began in the 1970s.

Temperatures cooled slightly in the 1960s and 1970s from the temperatures of the 1930s and 40s. Climatologists told the world that carbon dioxide was responsible for this *cooling*. On Earth Day of 1970 Dr. Kenneth Watt warned, "The world has been chilling sharply for about twenty years. If present trends continue, the world will be about four degrees colder for the global mean temperature in 1990, **but eleven degrees colder in the year 2000.** This is about twice what it would take to put us into an ice age."[1] (emphasis added)

A *Washington Post* article published in 1971 warned, **"In the next 50 years [2021] fine dust that humans discharge into the atmosphere by burning fossil fuel will screen out so much of the sun's rays that the earth's average temperature could fall by six degrees.** Sustained emissions over five to ten years, could be sufficient to trigger an ice age."[2] (emphasis added)

Newsweek Magazine declared in April of 1975, "There are ominous signs that Earth's weather patterns have begun to change dramatically...The evidence in support of these predictions has now begun to accumulate so massively that meteorologists are hard-pressed to keep up with it...The central fact is that...the earth's climate seems to be cooling down...If the climate change is as profound as some of the pessimists fear, the results could be catastrophic."

The decade of the 1930s to 1940s was the warmest in recent experience. By the 1960s and 1970s, it was evident that temperatures were decreasing even though atmospheric CO_2 was rising at the same time. If CO_2, a product of combustion, causes global warming, we have to wonder how the earth could have been cooling at the same time and why they were blaming the same gas — carbon dioxide — for this cool down as they did for the slight warming of 1980–1998.

In June of 1975, Nigel Calder, writing in *International Wildlife*, stated, "The continued rapid cooling of the earth since WWII is in accord with the increase in global air pollution associated with industrialization, mechanization, urbanization and exploding population. Articles about global cooling abounded in magazines, newspapers, TV reports and books. Some of the suggested solutions to this "crisis" included taking away our rights and freedoms. For example, Arnold Reitze, a professor at Case Western Reserve University, warned, "We will be forced to sacrifice democracy by the laws that will protect us from further pollution." He suggested outlawing internal combustion engines, placing rigid controls on all research and development in the production and marketing of new products and population controls, with a penalty imposed on couples producing more than the prescribed number of children."[3]

In his 1976 book, *The Cooling*, Lowell Ponte declared, "This cooling has already killed hundreds of thousands of people. If it continues and no strong action is taken, it will cause world famine, world chaos and world war, **and this could all come about before the year 2000.**" (emphasis added) Today's global warming alarmists are hoping that nobody remembers these dire predictions that were big news for nearly a decade. Trolls on social media are now trying to convince us that this cooling scare never happened.

When the temperatures started to rise again in the early 1980s, the alarmists changed gears and — in lockstep — started issuing opposite predictions: that the earth was about to warm uncontrollably, with anthropogenic [human-caused] carbon dioxide still being the culprit.

Two Decades, Two Messages

In 1982 Mostafa Tolba, executive director of the UN Environment Program (UNEP), stated, "The world faces an ecological disaster as final as nuclear war within a couple of decades unless governments act now." He predicted that if the world failed to act immediately, there would be "by the turn of the century, an environmental catastrophe which will witness devastation as complete, as irreversible as any nuclear holocaust."[4]

In a 1988 interview James Hansen of NASA's Goddard Institute for Space Studies (GISS) was asked how he thought the area around his office in New York would be affected in 20 years. [2008].

James Hansen of GISS

Hansen replied: "The West Side Highway [which runs along the Hudson River] will be under water. And there will be tape across the windows across the street because of high winds. And the same birds won't be there. The trees in the median strip will change...There will be more police cars...[since] you know what happens to crime when the heat goes up." [5]

These days Dr. Hansen is issuing equally stunning predictions. In a 2016 paper he stated: "The current rate of global warming could raise sea levels by 'several meters'[6] [21 feet] over the coming century, rendering most of the world's coastal cities uninhabitable and helping unleash devastating storms...Global warming of 2°C above pre-industrial times — the world is already halfway to this mark — would be "dangerous" and risk submerging cities."[7]

On June 30, 1989, the Associated Press released the following misinformation: "U.N. Official Predicts Disaster, Says Greenhouse Effect Could Wipe Some Nations Off Map. Entire nations could be wiped off the face of the earth by rising sea levels if global warming is not reversed **by the year 2000**. Coastal flooding and crop failures would create an exodus of eco-refugees, threatening political chaos." (emphasis added)

Brown, director of the New York office of the U.N. Environment Program, added that governments had a 10-year window of opportunity [till 1999] to solve the climate crisis.[8] Yet so far no countries have suffered any of the aforementioned damage, or been "wiped off the map".

A Sept. 19, 1989, article in the *St. Louis Post-Dispatch* stated, "New York will probably be like Florida 15 years from now." That would have been 2004.

On December 5, 1989, the *Dallas Morning News*, stated: "Some predictions for the next decade [1989-1999] are not difficult to make...Americans may see the '80s migration to the Sun Belt reverse as a global warming trend rekindles interest in cooler climates."

In 1990 Michael Oppenheimer of The Environmental Defense Fund made the following ominous prediction: "By 1995, the greenhouse effect would be desolating the heartlands of North America and Eurasia with horrific drought, causing crop failures and food riots..."[By 1996] "The Platte River of Nebraska would be dry, while a continent-wide black blizzard of prairie topsoil will stop traffic on interstates, strip paint from houses and shut down computers...The Mexican police will round up illegal American migrants surging into Mexico seeking work as field hands."[9] None of this ever happened, of course.

In 1990 Meryl Streep hosted a PBS 10-episode documentary entitled, "Race to Save the Planet". This documentary series predicted that by 2000 the world's average temperature would be four degrees warmer.

In January of 2000 Dr. Michael Oppenheimer of the Environmental Defense Fund, commenting in a NY Times interview on the mild winters in New York City, exclaimed: "But it does not take a scientist to size up the effects of snowless winters on the children too young to remember the record-setting blizzards of

1996. For them, the pleasures of sledding and snowball fights are as out-of-date as hoop-rolling, and the delight of a snow day off from school is unknown."[10]

The Big 2017 New York Snow

This was also totally incorrect. In fact, New York City experienced several significant snowfalls during the winter of 2017-2018. And on March 12, 2018, a blizzard with gale winds up to 50 mph hit eastern North Carolina, several hundred miles south of New York. Parts of Kentucky received about a foot of snow in that same weather system.

On March 29, 2001, *CNN* stated: "In ten year's time [2011], most of the low-lying atolls surrounding Tuvalu's nine islands in the South Pacific Ocean will be submerged under water as global warming rises sea levels." Yet on Feb. 9, 2018, 17 years later, *The Daily Mail* reported that the land area of this chain of islands, comprised of coral atolls, is actually *growing in size*. A study conducted by the University of Auckland in New Zealand used aerial photographs as well as satellite imagery to demonstrate that eight of the atolls and nearly three fourths of the islands grew in size between 1971 and 2014.[11]In 2005, Andrew Simms, policy director of the New Economics Foundation, announced, "Scholars are predicting that 50 million people worldwide will be displaced **by 2010** because of rising sea levels, desertification, dried up aquifers, weather-induced flooding and other serious environmental changes." (emphasis added) Fifty million people! This simply did not happen.

In June, 2008, Ted Alvarez of *Backpacker Magazine Blogs* wrote: "You could potentially sail, kayak, or even swim to the North Pole by the end of the summer. Climate scientists say that the Arctic ice...is currently on track to melt sometime in 2008."[12] Later in that same year a Russian icebreaker was stranded for a week in the ice of the Northwest Passage.[13] The truth is that Arctic ice has been at approximately the same level for the past thirty years, and we have the records to prove it.

The 1970s cooling scare and the present warming scare share these three conclusions:

1) Mankind's activities are responsible.

2) We of the industrialized world must atone for our sins.

3) We have the ability to control the climate.

The fact that none of these predictions and tipping points has materialized invalidates the AGW hypothesis. After all, if a hypothesis or theory is correct, scientists should be able to make accurate predictions based on it. Astronomers can predict solar and lunar eclipses to the exact minute, because they understand the positions and motions of the sun, moon and planets.

"Climategate"

In 2009 1,079 emails between climate scientists at the University of East Anglia's Climatic Research Unit (CRU) in the United Kingdom and others working with them, were leaked. This small group of scientists had been the most influential in driving the worldwide alarm over global warming. The emails they exchanged revealed that they were engaged in unethical practices, such as tampering with data and keeping scientists who disagreed with them out of the peer review system.

One email reads, "I've just completed Mike's Nature trick of adding in the real temperatures to each series for the last 20 years [i.e. from 1981 onward]." This referred to a graph constructed by Michael Mann of Penn State University in Pennsylvania, which used conflicting information to exaggerate the threat they claimed existed from global warming.

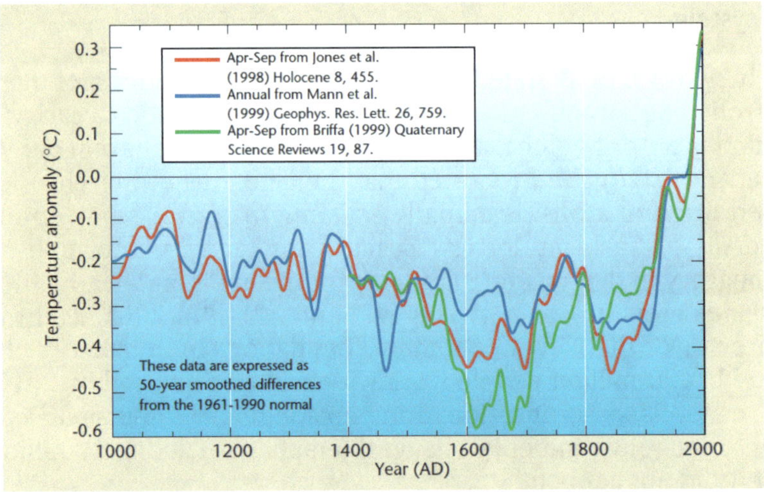

Mike Mann's Famous Hockey Stick Graph

The statistical methods Mann had used to create the uptick on the right side — which to some resembled a hockey stick — were first exposed as fundamentally flawed by Steve McIntyre, a Canadian statistician. A heated battle was raging between Mann's supporters and McIntyre over the statistical basis on which the CRU had constructed the global warming case for the Intergovernmental Panel on Climate Change (IPCC). Based on data taken from tree rings, this graph incorrectly showed that — after a thousand years of global temperatures being very steady — they had suddenly risen to their highest level in recorded history.

The resulting graph resembling a hockey stick was published everywhere, including the prestigious *Nature* magazine. But no one had bothered to check the data behind it. "Given star billing by the United Nations International Panel on Climate Change (IPCC), not least for the way it appeared to eliminate the long-accepted Medieval Warm Period when temperatures were higher than they are today, the graph became the central icon of the entire man-made global warming movement."[14]

However, that's not what the tree ring data showed. They actually indicated that temperatures had been declining. Because this wasn't what Mike Mann had anticipated, he decided to lop off the tree ring data from 1961 onward and replace it with data from thermometers. One leaked email stated, "Let's use Mike's trick to hide the decline."[15]

The leaked email documents also revealed discussions between these men about how they could manipulate data to get rid of the Medieval Warm Period, which lasted from about 950 AD to 1250 AD. Members of this elite group also considered tactics to prevent the release of the data on which the IPCC's conclusions were based: data requested by other scientists under Freedom of Information (FOI) laws. In Britain failure to release Freedom of Information requested data is a criminal offense.

Michael Mann, who created the Hockey Stick graph

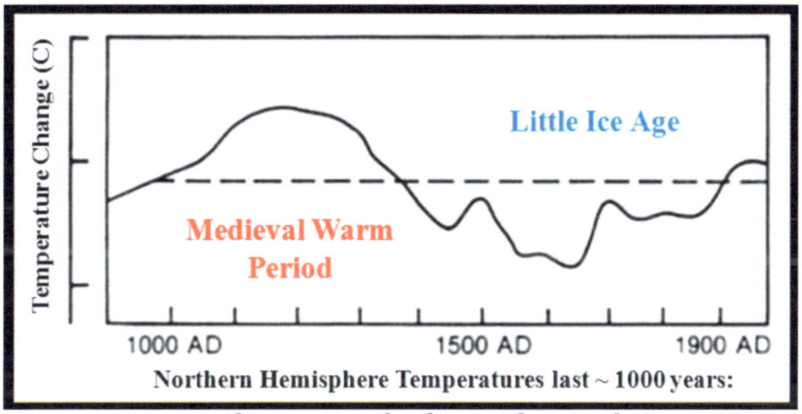

Correct Climate Graph of Last Thousand Years

Phil Jones, head of the Climate Research Unit, claimed to have lost large chunks of this data. Jones had commented, "I should warn you that some data we have we are not supposed to pass on to others. We can pass on the gridded data – which we do. Even if WMO agrees, I will still not pass on the data. We have 25 or so years invested in the work. Why should I make the data available to you, when your aim is to try and find something wrong with it?"[16]

This attitude violates the scientific method, which demands transparency. A defining characteristic of science is the sharing of data and procedures used in their study, so that independent parties can replicate and validate their observations and experiments. Scientists are expected to make requested data available to others who are endeavoring to understand how they reached their conclusions.

In their emails these scientists also expressed their frustration at the fact that temperatures had failed to rise as expected. One researcher lamented, "The fact is that we can't account for the lack of warming at the moment, and it is a travesty that we can't." If they were really worried that the earth was warming, shouldn't they have been elated that temperatures had leveled off?

Phil Jones, head of CRU

Equally shocking, these men revealed the ruthless methods they used to silence anyone who dared to question their conclusions. They had pressured editors of scientific journals to block the publication of studies contradicting their results. A professional scientist's career rests on his/her success in getting their studies published in peer reviewed journals. Phil Jones, writing to Michael Mann about two papers that disagreed with their hypothesis, stated, "Kevin and I will keep them out somehow – even if we have to redefine what the peer-review literature is!"

They discussed how they could pressure the editors of these journals not to publish studies that had reached conclusions differing from theirs. Mann wrote, "Perhaps we should encourage our colleagues in the climate research community to no longer submit to, or cite papers in, this journal?"

To this Jones responded, "I will be emailing the journal to tell them I'm having nothing more to do with it until they rid themselves of this troublesome editor."

They demonstrated their hatred of skeptical scientists, rejoicing at the news that one of them had died. Ben Santer, the same scientist who had reversed the conclusions of IPCC scientists in their first report in 1996, told CRU, "The next time I see Pat Michaels [a dissenting scientist] at a scientific meeting, I'll be tempted to beat the crap out of him."

Several emails indicated that they were trying to make the results of their research as sensational as possible, whether correct or not. On Sept. 19, 1996, Gary Funkhouser confessed, "I really wish I could be more positive about the...material, but I swear I pulled every trick out of my sleeve trying to milk something out of that....I don't think it'd be productive to try and juggle the chronology statistics any more than I already have — they just are what they are...I think I'll have to look for an option where I can let this little story go as it is."[17]

Geoff Jenkins' letter to Phil Jones on Nov. 22, 1996, reads in part, "Remember all the fun we had last year over 1995 global temperatures, with the early release of information (via Australia), 'inventing' the December monthly value, letters to Nature, etc., etc.? I think we should have a cunning plan about what to do this year, simply to avoid a lot of wasted time...We feed this selectively to Nick Nuttall (who has had this in the past and seems now to expect special treatment) so that he can write an article for the silly season...I know it sounds a bit cloak-and-dagger but it's just meant to save time in the long run."[18]

You may have wondered how the term "global warming" changed to "climate change". This was no accident. A 2004 email from Minns/Tyndall Center on the University of East Anglia campus stated, "In my experience with global warming, freezing is already a bit of a public relations problem with the media." To this came the reply from Bo Kjellen in Sweden: "I agree with Nick that climate change might be a better labeling than global warming."

A letter to the Norwich Green Party by Ian Harris, who worked for CRU, reveals the intense desire these scientists had to use scare tactics to force public policy: "We're looking at an unprecedented acceleration in temperature ... **Even if it turns out to be naturally-occurring, who's willing to take that chance? We should be trying to wean ourselves off of unsustainable energy generation**

and use anyway."(emphasis added) Harris was saying in effect that even if the warming should later be found to be from natural causes rather than from the actions of mankind, he didn't care.[19]

Christopher Booker has called this "the greatest scientific scandal of our age."[20]

After these emails were leaked, other climate scientists such as Dr. Eduardo Zorita and George Monbiot called for Dr. Jones to step down as head of the CRU. But he refused to do so.

James Delingpole, a journalist at the UK *Daily Telegraph*, described this collection of emails "conspiracy, collusion in exaggerating warming data, possible illegal destruction of embarrassing information, organized resistance to disclosure, manipulation of data, private admissions of flaws in their public claims and much more."

At first this caused quite a stir. And really, these leaked emails which revealed such dishonesty and arrogance should have marked the end of the climate alarmism fiasco. But unfortunately the major media covered up this affair, which should have been the biggest story of the year. And the Obama administration dismissed it as "irrelevant". So most people have no idea that this important information ever surfaced.

In order to deflect the blame from these corrupt scientists, several media outlets blamed the Russians for hacking their emails. "The computer hack, said a senior member of the Inter-governmental Panel on Climate Change, was not an amateur job, but a highly sophisticated, politically motivated operation. And others went further. The guiding hand behind the leaks, the allegation went, was that of the Russian secret services." James Delingpole called this attempt to deflect from the real story a "glorious red herring". They were likely released by a goodhearted soul working at that institution who had access and opportunity.

Other evidence of corruption within the climate alarmist circle has come to light. Testifying before a congressional committee on Dec. 6, 2006, Dr. David Deming, a University of Oklahoma professor, stated, "I received an astonishing email from a major researcher in the area of climate change. He said, 'We have to get rid of the Medieval Warm Period.'"[21]

Two years later, in 2011, 5,000 more emails were leaked, in what is now called Climategate II. These were even more startling and incriminating than the 1,079 released in 2009. Myron Ebell commented, "If there were any doubts remaining after reading the first Climategate emails, the new batch of emails that appeared on the web today [November 22] make it clear that the UN Intergovernmental Panel on Climate Change is an organized conspiracy dedicated to tricking the world into believing that global warming is a crisis that requires a drastic respo nse."[22]

In these emails Phil Jones, head of the CRU, indicated his continued efforts to keep important data under wraps. He said -

> "I've been told that IPCC is above national FOI Acts. One way to cover yourself and all those working on AR5 would be to delete all emails at the end of the process. Any work we have done in the past is done on the back of the research grants we get – and has to be well

> hidden. I've discussed this with the main funder [US Department of Energy] in the past, and they are happy about not releasing the original station data."

He also admitted that there were serious credibility problems with the models they were using. "Basic problem is that all models are wrong – not got enough middle and low level clouds...What he [Zwiers] has done comes to a different conclusion than Caspar and Gene! I reckon this can be saved by careful wording."

Another scientist truthfully pointed out, "It is inconceivable that policymakers will be willing to make billion and trillion-dollar decisions for adaptation to the projected regional climate change based on models that do not even describe and simulate the processes that are the building blocks of climate variability."[23]

One might think that, with these shady practices and double dealing revealed, the public would have demanded an end to the hoax. But again, this scandal was quickly disappeared from view as if it had never happened. Everything was carefully covered up, and most of the public still has no idea that any of these shenanigans had been revealed. You may never have heard about this smoking gun which proved that climate alarmism is built on lies.

---

1. www.aim.org/aim-column/washington-post-gives-megaphone-to-far-left-groups-
2. www.aim.org/aim-column/washington-post-gives-megaphone-to-far-left-groups-
3. "Cold Facts About Pollution," The Southeast Missourian, Jan. 29, 1970
4. Engdahl, op. cit
5. https://stevengoddard.wordpress.com/2013/03/10/another-hansen-fail/
6. "Several" could be five or more meters. Several meters equals about 17 feet, or from the ground to the roof of a house.
7. www.theguardian.com/science/2016/mar/22/sea-level-rise-james-hansen-climate-change-scientist
8. www.theintermountain.com/opinion/letters.../disaster-predictions-often-turn-out-false/
9. http://www.foxnews.com/tech/2010/12/30/botched-environmental-forecasts.html
10. http://www.climatedepot.com/2014/02/18/flashback-2000-new-york-times-sledding-and-snowball-fights-are-as-out-of-date-as-hoop-rolling/
11. http://www.dailymail.co.uk/sciencetech/article-5371535/Sinking-Pacific-nation-getting-bigger-study.html
12. https://thebigfeedblog.blogspot.com/2008/12/the-worst-climate-predictions-of-2008.html

13. http://losingamerica.truth-in-politics.info/?p=3142

14. Christopher Booker, www.telegraph.co.uk/comment/ columnists/ christopherbooker/6679082/Climate-change-this-is-the-worst-scien tific-scandal-of-our-generation.html, Nov. 28, 2009

15. "Climategate 'hide the decline' explained by Berkeley professor Richard A. Muller", YouTube Video posted on 2/27/11

16. Email to Warwick Hughes, Feb. 21, 2005

17. https://www.lavoisier.com.au/articles/greenhouse-science/climate-change/ climategate-emails.pdf

18. Ibid.

19. Costella, op. cit., p.26

20. Christopher Booker, "The Real Global Warming Disaster: Is the Obsession with 'Climate Change' Turning Out to be the Most Costly Scientific Blunder in History?", 2009

21. Tony Heller, "Erasing The Medieval Warm Period" https://real climatescien ce.com/2018/12/erasing-the-medieval-warm-period/, Dec. 23, 2018

22. Myron Ebell, Director of the Competitive Enterprise Institute's Center on Energy and Environment.

23. Larry Bell, "Climategate II: More Smoking Guns From The Global Warming

Chapter Three

The Role of "Greenhouse Gases"

> "The science is unequivocal...President Obama and I believe very deeply that we do not have time for a meeting anywhere of the Flat Earth Society." – John Kerry, Biden's Climate Czar

Carbon Dioxide and Water Vapor

On a sunny day the temperature inside your car is much warmer than it is outside. Just as in a greenhouse, sunlight streaming through the glass is trapped inside as heat. Greenhouses and automobiles, having walls and roofs, are closed systems. However, Earth is not a closed system. Its atmosphere has no walls or roof. All of the sunlight falling on Earth's surface is radiated back into space. If this weren't true, Earth would just keep getting warmer and warmer. So the term "greenhouse effect" is a misnomer.

A few atmospheric gases, such as water vapor, carbon dioxide and methane, hold onto some of this incoming energy for a short time before releasing it. These are called greenhouse gases, even though, not being a closed system, Earth isn't actually a greenhouse. Without these gases the earth's average temperature, about 15°C, (59°F.) would instead be 28 degrees colder, - 13°C. (close to zero Fahrenheit), and life as we know it couldn't exist. But a greenhouse effect that was out of control could theoretically raise temperatures higher than we'd want them to be.

Contrary to what we've been told, the main greenhouse gas is water vapor. Making up about 3% of Earth's atmosphere, *water vapor is at least seventy times as powerful in atmospheric warming as is carbon dioxide*, which makes up only about 0.04% of the atmosphere: just four hundredths of one percent, or 400 parts per million (ppm). In ideal conditions at the equator sunlight falling on one square meter generates approximately 1000 watts of power. According to Dr. Richard Lindzen, physics professor at the Massachusetts Institute of Technology, the radiative forcing from this amount of CO_2 makes a difference of only about 1 watt/square meter.

Most CO_2 in the atmosphere is emitted from the oceans and the tropics of South America and Africa, where the sunlight is more direct. Oceans cover about 70% of Earth's surface. CO_2 also comes from the earth's interior through volcanoes. The earth has over 1500 active volcanoes, many of which spew lava underwater through cracks where tectonic plates are separating. This carbon dioxide has been stored in the earth's mantle since our planet was formed billions of years ago.

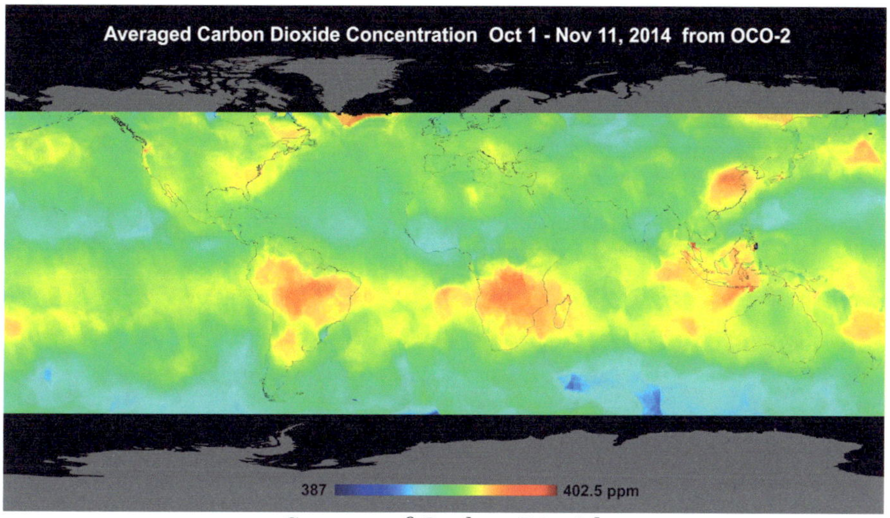

Sources of Carbon Dioxide

As you can see from this map, the red and bright yellow areas emitting the most CO_2 are in Africa, which has few industries or vehicle traffic, central South America, which is largely unpopulated, the tropical oceans and southeastern China, which is heavily populated and industrialized. There is much less CO_2 being emitted in the United States, western Europe and Russia.

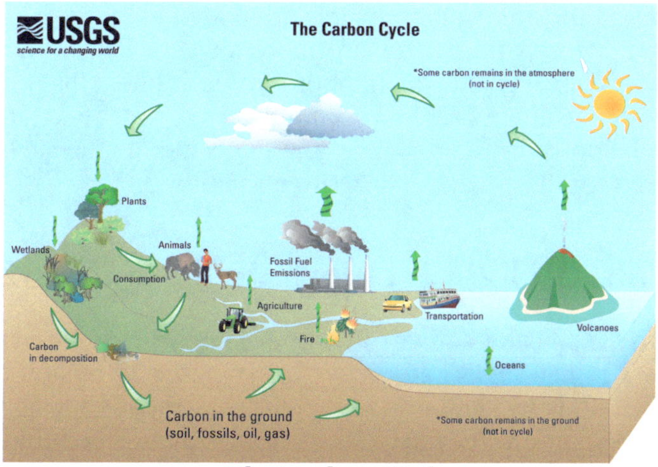

The Carbon Cycle

Carbon dioxide is also emitted by dead decaying organisms, termites, bogs and swamps. This exchange between plants, animals, rocks, the atmosphere and the oceans is known as the carbon cycle. A study reported by the New York Times concluded that termites, digesting vegetable matter with the help of bacteria, emit more than twice as much carbon dioxide as all the world's smokestacks.[1] Termite activity has been particularly high due to the widespread clearing of land for raising corn and sugar cane to manufacture ethanol and for building wind and solar farms. These operations have provided abundant food for them in the leftover debris.

We've been told that when carbon dioxide is released, it moves up into the atmosphere. But this is scientifically impossible. Because carbon dioxide is more dense than oxygen and nitrogen, it sinks downward to provide food for plants (and us) through the process of photosynthesis. Photosynthesis is the most important chemical reaction on earth, because it provides all of our food. Plants use it to make food for themselves. Animals then eat the plants, absorbing these nutrients. Notice the zigzag structure of this line in this graph extending from 1960 to 2016. The reason for that is that every year carbon dioxide is more prevalent in the winter. Plants absorb it in the summer, which lowers the amount recorded just a bit.

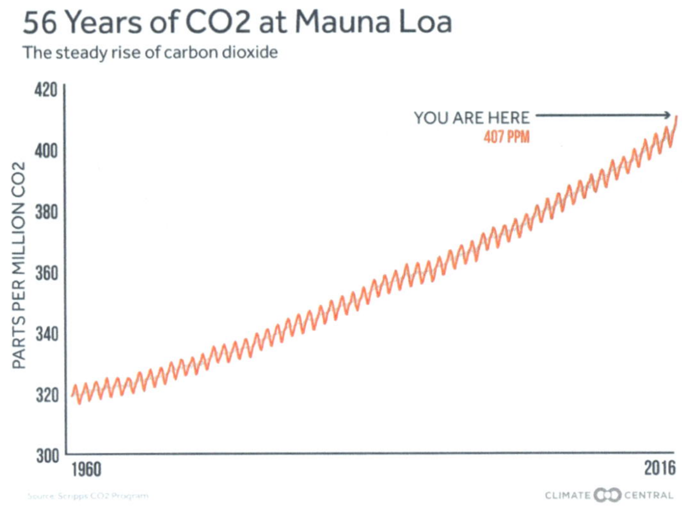

Carbon dioxide is rising steadily, not exponentially.

With more CO_2 in the atmosphere, plants are conducting more photosynthesis, producing more carbohydrates and also more oxygen. This is an added benefit to the earth that is never discussed. In photosynthesis (meaning "putting together with light") Six molecules of CO_2 and six molecules of water react in the presence of sunlight and chlorophyll to produce one molecule of glucose [sugar] and six molecules of oxygen. The reaction is written as follows.

$$6CO_2 + 6H_2O + \text{light \& chlorophyll} \longrightarrow C_6H_{12}O_6 + 6O_2$$

CO_2 seeps into the soil and bodies of water, joining with other elements such as calcium to form animal shells and rocks. An estimated 39 million gigatons (3.9×10^{15} tons) of carbon is locked up in the shells of ancient animals whose bodies make up the ocean floor. Contrary to another lie we've been told, CO_2 doesn't remain in the atmosphere for long periods of time. The average carbon dioxide molecule stays in the atmosphere for only about fifteen days.[2]

Increasing the amount of CO_2 in the atmosphere from 0 to 280 ppm [parts per million] increases the atmospheric opacity [absorption] about 6.2%. Increasing it from 280 ppm to 407 ppm (where we are now) makes a difference of only 0.6%, one-tenth as much. This is because carbon dioxide absorbs and emits energy *within a very limited frequency range.* It briefly holds photons of light at just two wavelengths: 4.26 micrometers (asymmetric stretching mode) and 14.99 micrometers (bending vibrational mode). Once those two wavelengths have been saturated, further additions of CO_2 to the atmosphere make virtually no difference. Dr. Will Happer, physicist and Professor Emeritus at Princeton University, equated this to painting a barn red. The first two coats of paint change the color to red. But the next ten coats would not be noticed.

This graph shows the relatively small effect of carbon dioxide compared to water vapor and the truly minuscule effect of methane and nitrous oxide (laughing gas) on the atmosphere. As you can see, most of the warming is from water vapor. Compared to water vapor and carbon dioxide, because they exist in such minute quantities, methane and nitrous oxide have an almost zero effect. This important graph explains the effect of various greenhouse gases.

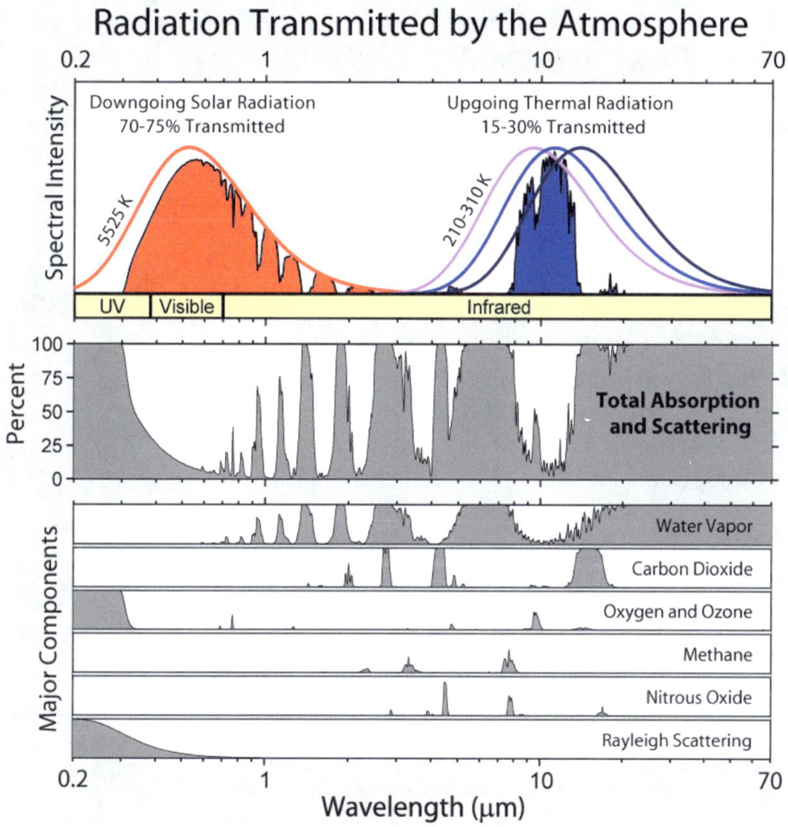

Atmospheric "Greenhouse Gases"

Yet, despite this undeniable fact, the World Economic Forum (WEF) is putting farmers all over the world under intense pressure to reduce the amount of methane and nitrous oxide produced on their farms. Using this excuse the government of the Netherlands has stolen thousands of farms. This is adding to the impending food shortages, which could possibly cause millions of people to starve.

A scientist at the US Department of Agriculture wrote in 1941:

> "Much has been written about varying amounts of carbon dioxide in the atmosphere as a possible cause of glacial periods. The theory received a fatal blow when it was realized that CO_2 is very selective as to the wavelengths of radiant energy it will absorb, filtering out only such waves as even very minute quantities of water vapor dispose of anyway. No probable increase of atmospheric CO_2 could materially

> affect either the amount of insolation [sunlight] reaching the surface or the amount of terrestrial radiation lost to space."[3]

Even if CO_2 were to rise to 690 ppm, the atmospheric opacity it causes would increase by only another 1%.[4]

The earth has become noticeably greener as a result of the slight increase in carbon dioxide. Compared to a 100-yard-long football field, this increase in carbon dioxide would amount to just three-eighths of an inch. This small increase in carbon dioxide has been a boon to agricultural production, which has increased about 15%. Because there is more carbon dioxide, plants don't need to keep their pores open as much to take it in. This saves water, which evaporates through the tiny pores. They grow faster, absorbing more CO_2 in the process. This is a buffering effect or negative feedback mechanism. The earth is noticeably greener, evidenced in satellite photos.

Yet in December of 2009 the Environmental Protection Agency (EPA) under the Obama administration erroneously classified this life-giving compound as a "pollutant".[5] Since then alarmists have been using the phrase "carbon pollution" to describe the release of carbon dioxide into the atmosphere.

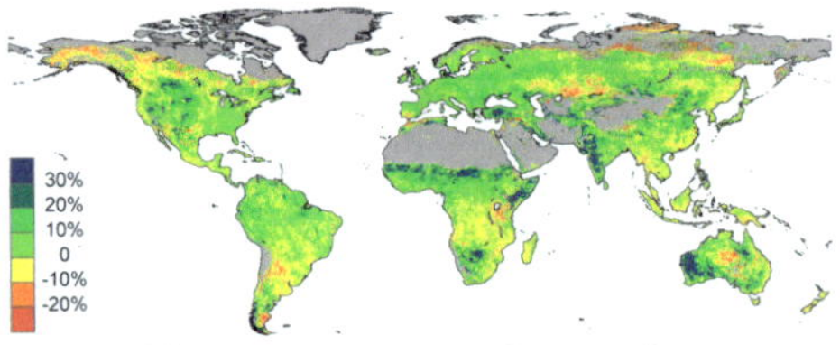

Satellite composite photo of greening earth. Courtesy of NoTricksZone.com

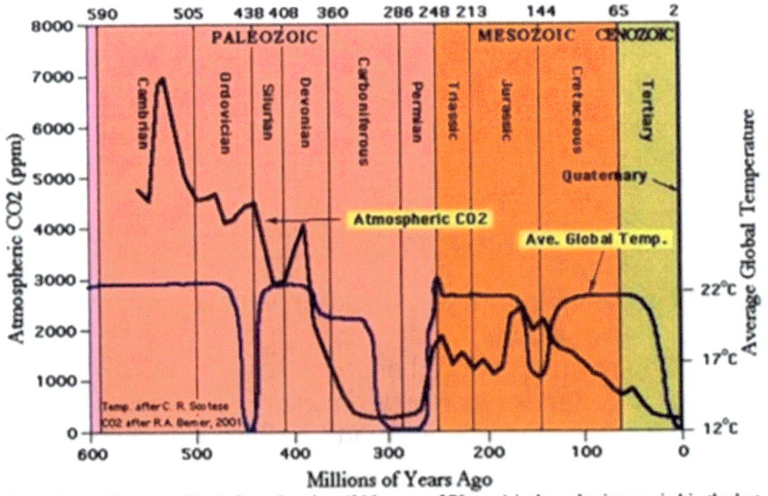

Late Carboniferous to **Early Permian** time (315 mya -- 270 mya) is the only time period in the last 600 million years when **both** atmospheric **CO2** and **temperatures** were as low as they are today **(Quaternary Period).**
Temperature after C.R. Scotese
CO2 after R.A. Berner, 2001 (GEOCARB III)

Quora.com

This graph depicts the levels of CO_2 and temperature for the past six hundred million years. As you can see, there have been times — such as 440 million years ago — when the earth was in an ice age, and the percentage of CO_2 was ten times higher than now. For most of this time the earth's average temperature has been higher than it is now. This graph demonstrates that on a large time scale there is no correlation between temperature and carbon dioxide. For the last 250 million years the average amount of carbon dioxide in the atmosphere has been about 1200 ppm, the optimum level for plants.

The photos of ominous looking clouds rising from the chimneys of coal burning plants, portrayed in the media, are actually clouds, not carbon dioxide. Whereas you can see clouds — made of tiny drops of liquid water — carbon dioxide is an *invisible* gas which can't be seen in photographs.These misleading photos are sometimes even touched up on Photoshop to increase contrast, making them appear menacing. Some people have also confused life-giving carbon dioxide (CO_2) with carbon monoxide (CO), an extremely toxic gas. The source of these clouds is combustion, of which water is a product.

Water droplets (clouds) Carbon dioxide is invisible.

The earth's oceans, covering about 70% of our planet, contain about 40,000 gigatons [billion tons] of CO_2. This is 50 times as much as the atmosphere holds: 800 gigatons. When ocean water warms, some of this carbon dioxide comes out of solution. This is because more carbon dioxide can dissolve in cold water than in warm water. The same thing happens when you let a cold glass of soda pop sit until it reaches room temperature. The carbon dioxide that made it fizzy escapes as the liquid warms up. Another glass of pop left in the fridge keeps its fizz for a longer time.

Because oceans have 250 times the mass of the atmosphere, they have a significant effect on climate. Water has a high specific heat. This means that it takes much more energy to change its temperature than the temperature of the land. So oceans hold a thousand times more heat than either the land or the atmosphere.

We've been told repeatedly by the media that because of increased CO_2 the oceans are becoming more acidic. This made-up story is also scientifically impossible. Carbon dioxide reacts with water to form carbonic acid, a weak acid [H_2CO_3]. When CO_2 is released from the ocean into the atmosphere, there's less, not more, carbon dioxide - and carbonic acid - in the water. So when ocean water warms, releasing CO_2 into the atmosphere, it becomes less — not more — acidic.

Neutral liquids such as water, containing equal amounts of H+ and OH- ions, have a pH of 7.0 on the pH (potential hydrogen) scale. Acids have a pH of less than 7.0. Bases have a pH that's higher than 7.0. The pH of ocean water ranges from 7.5 to 8.3. Ocean water is basic, not acidic.

Because water at different latitudes heats unequally, due to the angle at which sunlight strikes it, ocean currents are constantly circulating. Cold, dense water sinks and warmer water rises. This is not to say that a tidal pool into which acidic water from the land is flowing, could have a slightly lower pH. But, in general, ocean water could never become even close to acidic.[6]

Dr. Patrick Moore, one of the founders of Greenpeace, an organization that was responsible for stopping nuclear testing in the atmosphere, and also the hunting of whales and baby harp seals, came to a very interesting conclusion about the percentage of CO_2 in the atmosphere.

[7]"Human emissions of carbon dioxide have saved life on Earth from inevitable starvation and extinction [at 150 ppm] due to lack of CO_2. To use the analogy of the Atomic Clock, if the Earth were 24 hours old, we were at 38 seconds to midnight when we reversed the trend towards the End Times. If that isn't good news I don't know what is. You don't get to stave off Armageddon every day.

"I issue a challenge to anyone to provide a compelling argument that counters my analysis of the historical record and the prediction of starvation based on the 150-million-year trend...I submit that much of society has been collectively misled into believing that global CO_2 and temperature are too high when the opposite is true for both. Does anyone deny that below 150 ppm CO_2 plants will die? Does anyone deny that the Earth has been in a 50-million-year cooling period and that this Pleistocene Ice Age is one of the coldest periods in the history of the planet?"[8]

Dr. Moore points out that the percentage of carbon dioxide in the atmosphere had been declining for the last 150 million years: from 6000 ppm [parts per million] to just 180 ppm. Since plants *must* have at least 150 ppm of CO_2 to survive, they will die if the percentage is lower than this. Since plants also feed animals, we are talking about not just plants but ***all*** life on earth dying out should the percentage of carbon dioxide dip to a lower level than 150 parts per million.

Methane [CH_4] has also received a bad rap from the climate alarmists. One article states: "Due to the chemical bonds within its molecule, methane is much more efficient at absorbing heat than carbon dioxide...making it a very potent greenhouse gas. Fortunately, methane can only last about 10 to 12 years in the atmosphere before it is oxidized and turns into water and carbon dioxide."[9]

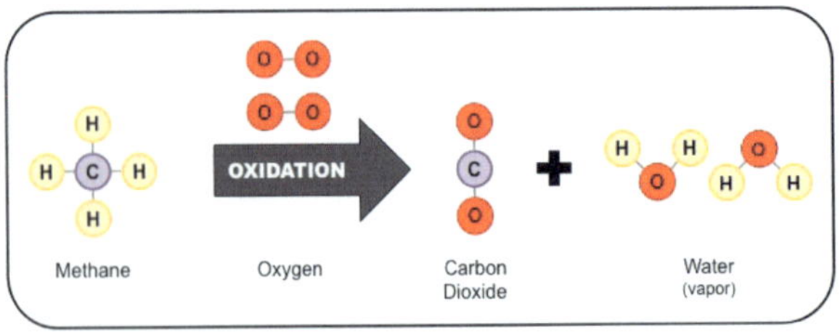

This is also incorrect. Methane is oxidized by oxygen, but because the atmosphere contains so much oxygen, the process happens much faster than 10-12 years. Each molecule of methane, or natural gas, has more efficient heat trapping qualities than CO_2. So now they are pegging it as a "dangerous" gas. In 1995 the IPCC declared it to be 21 times as strong as carbon dioxide. Yet methane makes up just 0.00002% of the atmosphere — totally insignificant — meaning that this is just another huge scare tactic.

Methane and nitrous oxide have been demonized, because the alarmists want to see the world's people get rid of all their ruminant farm animals — cows, goats and sheep — that they say are burping it out: 3.5 billion of them. Because methane is also produced in the growing of rice, the policy makers with this agenda say that farmers will no longer be permitted to grow this crop that is a staple of half of the world's people.

Arctic peat bogs that are frozen over in the winter are one of the largest natural sources of methane. Anaerobic degradation of submerged Sphagnum species — mosses growing in peat bogs, produces significant quantities of methane. However, a study on peat mosses in the Netherlands revealed that a large percentage of this methane is being eaten by methanotrophs, bacteria that oxidize methane.[10]

The US and the European Union pressured over 100 nations to pledge to reduce the amount of methane and nitrous oxide in their countries a whopping 30% by the year 2030. How will they do that unless they kill the farm animals and decrease the amount of rice grown? If these nations actually go through with this, milllions of people will starve in the near future. This is already happening in Sri Lanka, Ghana and South Africa. Despite the leaks of methane and despite the fact that farm animals breathe it out, its existence is still only two parts per million, compared to over 400 parts per million of carbon dioxide.

---

1. https://www.nytimes.com/1982/10/31/us/termite-gas-exceeds-smokestack-pollution.html
2. Dr. Fred Goldberg lecture, (YouTube video) March 15, 2016
3. Prof. Murry Salby, "Climate Scientist Murry Salby Demolishes the Global Warming Alarm," lecture on YouTube, Aug. 13, 2014
4. https://stevengoddard.wordpress.com/2016/04/17/global-warming-scam-has-been-dead-for-75-years/
5. http://dailycaller.com/2015/02/11/epa-challenged-carbon-dioxide/
6. Dr. Don Easterbrook, lecture before Washington state senate committee, March 26, 2013
7. Dr. Robert Giegengack, former chair, Dept. of Earth and Environmental Science, University of Pennsylvania
8. https://www.technocracy.news/former-president-of-greenpeace-scientifically-rips-climate-change-to-shreds/
9. Frederick Beaudry, "Methane, a Powerful Greenhouse Gas," Treehugger,11/25/19
10. Nardy Kip, et al, https://www.nature. com/articles/ngeo939/

Chapter Four

False Climate Change Claims

"The truth has no defense against a fool determined to believe a lie."
- Mark Twain, author

"Polar bears face extinction due to the melting Arctic."

The headline of an undated article on the World Wildlife Fund website reads, "Polar Bear Population Decline a Wakeup Call for Climate Change Action."[1] In reading this article we learn that it is the polar bear population in the Beaufort Sea — not the world's polar bear population — that has declined, another example of misrepresenting information.

In contrast, a 2011 article in International Business Times stated, "The US Fish and Wildlife Service estimates that the polar bear population is currently at 20,000 to 25,000 bears, up from as low as 5,000-10,000 bears in the 1950s and 1960s.[2]The current polar bear population is estimated at between 22,000 and 31,000 bears. [3]This number is verified by the IUCN/SSC Polar Bear Specialist Group.[4] Because of strict hunting regulations, the number of polar bears has been increasing for many years.

Dr. Susan Crockford, one of the world's premier experts on polar bears, reported that

> "Polar bears turned out to be more flexible and resilient than predicted and many subpopulations are better off than before...There is no existential emergency for polar bears or any other Arctic sea mammals due to declining summer sea ice, despite continued messages of doom from remorseless experts."[5]

So we can put our fears to rest that polar bears are declining in number.

"97% of scientists say that human-caused climate change is a serious threat."

We are repeatedly told that a 97% consensus [agreement] of scientists say that Earth's climate is experiencing dangerous warming caused by human activities. This erroneous and totally misleading figure was obtained from four different studies, all of which were flawed.

The first was by Naomi Oreskes. The topics of many articles counted in her study failed to address the causes of climate change, but — assuming that IPCC conclusions were correct — instead focused on the influence of climate change on the incidence of influenza, the life cycle of frogs, etc. Before publishing Ms Oreskes' article, the journal editor failed to check either her methodology or her sources.

In another study, conducted in 2009, Maggie Zimmerman, a graduate student, sent a 2-minute online survey to 10,257 employees of schools and government research agencies. This left out over 10,000 geologists, physicists, meteorologists and astronomers who were studying the issue. Although she received 3,146 responses, her 98% figure was based on only 79 responses that she had hand picked. This was certainly not a representative sample.

In 2010 another college student, William Anderegg, identified 908 scientists who had written the most papers about global warming. He considered these to be most qualified to hold an opinion on climate change. "The 50 most prolific alarmists were published an average of 408 times each, versus only 89 times each for the skeptics." Hundreds of the studies he considered had several — as many as twelve — different authors. This misinformation weighted the results against the skeptics.[6] Anderegg failed to consider that the quantity of published studies doesn't equal correctness.[7] Scientists whose studies reach conclusions agreeing with the official government position on climate change are given lucrative grants. In contrast, those scientists who study the *natural* causes of climate change must scramble for financing. This is one reason why these scientists aren't publishing as many peer-reviewed papers.

In the fourth study John Cook, a professional cartoonist, issued a report to the Global Warming Policy Foundation stating that 97.1% of the 11,944 scientific papers he examined "explicitly or implicitly suggested that human activity was responsible for some warming."[8] This claim has been debunked by several other studies. A study by Legates found that only 41 of these 11,944 papers (or 0.3%) stated that most of the warming since 1950 was man made.[9]

Cook himself admitted that about two-thirds of the abstracts [study summaries] he looked at expressed no opinion about man made global warming. These were excluded from the results. This action reveals the dishonest character of the study. Laughably, Cook had told the publisher of his study its results (97% consensus) before he even did the research.[10] Regarding COVID-19, one person quipped, "90% of scientists agree with the people who fund them." This is also obviously true of climate scientists. It's human nature.

But science doesn't operate by consensus. To adhere to the scientific method, scientists are obligated to question all new ideas. In that case, even if this 97% figure was correct (it isn't), it would be meaningless. When Galileo challenged the scientific consensus that the earth was the center of the universe, he was met by an extremely hostile reaction.

On the other hand, in the 2007 Global Warming Petition project 31,487 scientists, over 9,000 of whom had PhD's, signed the following declaration:

> "There is no convincing scientific evidence that human release of carbon dioxide, methane or other greenhouse gases is causing, or will, in the foreseeable future, cause catastrophic heating of the Earth's atmosphere and disruption of the Earth's climate. Moreover, there is substantial scientific evidence that increases in atmospheric carbon dioxide produce many beneficial effects upon the natural plant and animal environments of the earth."[11]

A prominent scientist, Hal Lewis, resigned from the American Physical Society (APS) in 2010 after noting that this society had been taken over by unscientific global warming hysteria. In his letter of resignation, he stated: "The global warming scam, with the (literally) trillions of dollars driving it, that has corrupted so many scientists, has carried APS before it like a rogue wave. It is the greatest and most successful fraud I have seen in my long life as a physicist."[12]

Many scientists have been afraid to disagree with those in power out of fear that their careers will be ruined. Since the global warming hypothesis surfaced about 40 years ago, many meteorologists and climatologists have waited until they retired to speak up. Others became skeptical after more information surfaced. Just a few of the top scientists who are now standing up and declaring that this is wrong are: Dr. David Evans, Dr. Joanna Simpson, Dr. Judith Curry, Dr. Alrude Allegre, Dr. Tad Murty, Dr. Chris de Freitas, Dr. Kiminori Itoh, and Dr. Andrei Kapitsa.[13]You need to Google the names of these scientists to learn why they've changed their stance.

Scientists choosing to investigate natural rather than man-made causes of warming are denied research grants. This is obviously true, because government officials, lusting for carbon taxes, are interested only in those reports that can show in some way that humans are causing this "problem". For example, if a scientist can demonstrate that a species of butterfly has migrated north or is in danger of extinction, attributing this to human-caused climate change will guarantee them lucrative grants that are perpetually renewed. Scientists are human beings, and they don't want to have their careers ruined by losing their government grants. They know these grants will be renewed as long as there is a "problem".

Scientists who speak out are drowned out by the tsunami of voices of climate hysteria. On Nov. 29, 2012, 125 scientists wrote an open letter to the United Nations [14]declaring that there is no climate emergency. This was ignored. Then on Sept. 23, 2019, the same day Greta Thunberg made her speech at the UN, 500 prominent scientists sent a letter to the international body. It stated:

> "The general circulation models of climate on which international policy is at present founded are unfit for their purpose. Therefore, it is cruel as well as imprudent to advocate the squandering of trillions of dollars on the basis of results from such immature models. Current climate policies pointlessly and grievously undermine the economic system, putting lives at risk in countries denied access to affordable, reliable electrical energy."

"Ice in the polar regions is melting as never before."

To depict an impending catastrophe, the first scene of a 2004 science fiction movie, "The Day After Tomorrow", showed a coastal portion of the West Antarctic ice sheet crashing into the ocean. But recently, after detecting unexplained seismic activity in that region, geologists from Washington University in St. Louis accidentally discovered the cause of that warming: a chain of about 90 active volcanoes hidden under this glacier.[15]

According to NASA, the total amount of ice in Antarctica has been increasing for several decades.[16] The Antarctic glacier, containing about 90% of the earth's fresh water, is exceptionally stable. It is almost three miles thick (16,400 feet) in the center. NASA stated, "The Antarctic ice sheet showed a net gain of 112 billion tons of ice per year from 1992 to 2001."[17]

But despite these earlier gains it was recently reported that there has been a steep downturn in Antarctic ice. "Scientists wonder if it's 'the beginning of the end'"[18]With their usual flair for exaggeration, the doomsayers claim that the Antarctic is heating up out of control. But satellite photos of the continent show much more sea ice around Antarctica than existed in a 1976 photo in *National Geographic.* In 2020 *CNN* reported that Antarctica had the coldest temperatures on record.[19]This graph shows the temperature trend since 1980.

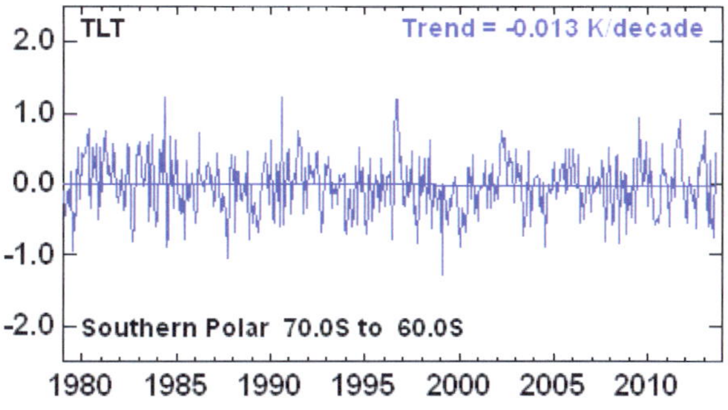

Antarctic Temperatures for the past 40 years

The average temperature in Antarctica is -58° F. To get the ice to melt, you'd have to raise that temperature about 90° to water's melting point of 32°F. That's a huge increase in temperature. There have been no gaps in the Antarctic ice core record for the last 15 million years.[20] The snow and ice at the South Pole has been accumulating so rapidly that the first two scientific observation stations had to be abandoned after being taken over by ice. The third station was constructed on stilts so that it could be raised as needed.[21]

But what about the Arctic? Haven't we been told that the Arctic ice is melting? Well, a plane that crash landed on the ice in Greenland in 1942 was recovered fifty years later in 1992 under 268 feet of ice. This demonstrates that Greenland's ice has been getting thicker, not thinner. The Greenland glacier is melting at the edges but getting thicker in the interior. Of course, every summer there is some melting of icebergs near the ocean. The following winter that melted ice is replaced by more snowfall. This has been occurring for thousands of years.

National Geographic magazine predicted that all of the snow and ice in the Arctic would be melted by 2008. Canada.com predicted this would happen between 2010 and 2015. A leading expert at NASA predicted that all the snow and ice in the Arctic would be gone by 2012. "The Arctic is screaming," declared Mark Serreze, senior scientist at the national snow and ice data center in Boulder, Colorado. [22]The BBC put the year for disappearing Arctic ice at 2013. Nobel prize winner, Al Gore, predicted that all the ice would be gone in 2014. Being more conservative, James Hanson of NASA's Goddard Space Center named 2018 as the year for this occurrence. But, as you can see from this graph, in 2022 the average Arctic ice extent was near the top of the average number of square miles for the period of 1981 to 2010.

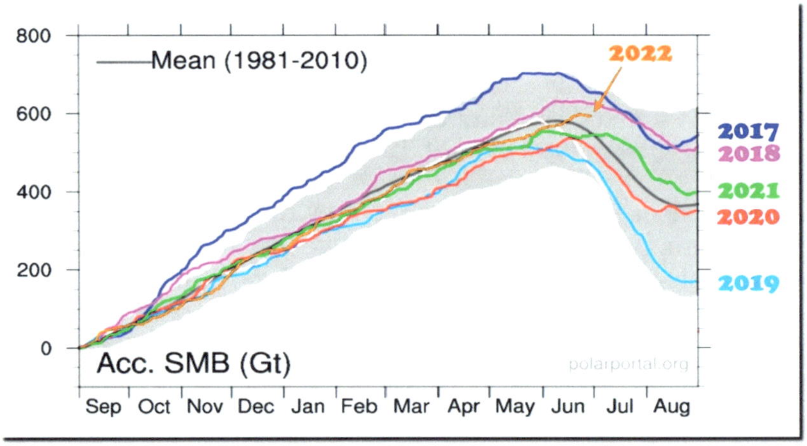

Arctic Ice Extent for DIfferent Years and Months of the Year

A large Greenland glacier, named Jakobshavn, which was thinning out around 2012, started growing again in 2017, according to NASA's Oceans Melting Greenland (OMG) project. The study's lead author, Ala Khazendar, a NASA glacialist on the OMG project, believes that the cooling of the North Atlantic due to the shift in the Atlantic Multidecadal Oscillation probably caused the glacier to rebound. Other NASA scientists have called this a "miracle".[23]The Atlantic Multidecadal Oscillation, a repeating cycle of changing ocean currents in the Atlantic Ocean which lasts for 50-70 years, influences the temperature of Greenland and other Arctic areas.

The North Pole is covered by sea ice averaging about three meters in thickness. There have been warmer times in recorded history when ships were able to navigate through the Northwest Passage. This isn't possible at the present time.

For generations alarmists have been predicting that Arctic ice would soon disappear. Newspaper and magazine articles in the early 1900s also made that prediction.[24] From 1880 to 1915 glaciers advanced. Then they receded during the period from 1915 to 1945, a period of warming. They started advancing again from 1945 to 1977 when Earth was again cooling.

A good example of a half-truth so common in our media is the following statement: "Melting in the Arctic reached an all-time high in June. Ice has been disappearing at 29,000 square miles a day."[25] This turns out to be another good way of lying with statistics. Dr. Tim Ball explained what is really happening here: "This is near the average daily rate of melt in the brief Arctic summer, but few people know this is natural. Approximately ten million km^2 melts every summer in approximately 145 days."[26] Usually all or more of this melted ice is replaced the following winter.

The red trend line in the following graph shows that the extent of ice in the Arctic has been almost perfectly constant during the period from 2006 to the present time. The blue line moving up and down reflects the freezing and thawing in response to summer and winter temperatures. There was a slight reduction in the amount of ice in 2012, but it quickly rebounded. The MASIE-NH Daily Image Viewer, published by NOAA, shows the amount of snow and ice in this area at any particular time. It can be accessed by clicking the link: https://nsidc.org/data/masie.

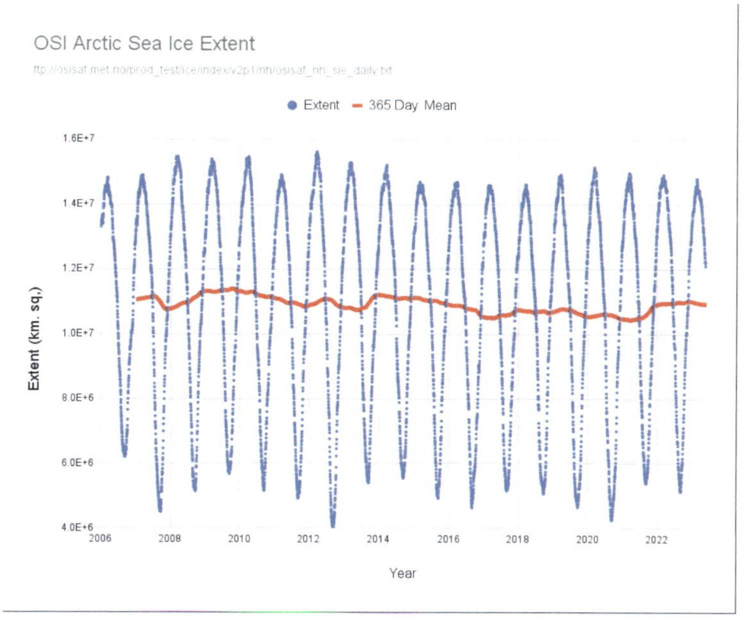

"Although the idea that a solid ice sheet covers the central Arctic has lingered stubbornly in the popular fancy, the northern cap of ice worn by our planet is actually a thin crust— on the whole, only about seven feet thick." [27]

"Sea Level Rise is Accelerating."

For the last 8,000 years the earth's sea level has been slowly rising at a constant rate. Nils Axel Mörner, a world-renowned expert on global sea level measurement, warned the UN Intergovernmental Panel on Climate Change (IPCC) that it was publishing false information that would inevitably be found to be incorrect. When they ignored him, he resigned. Dr. Mörner has affirmed that Earth's climate is cooling and that ocean levels will decrease. He reports that the sea level has been rising at an average rate of only 1.1 mm – about the thickness of your fingernail – per year. This figure is confirmed by NOAA.[28] This sea level increase adds up to about seven inches per century, not the 1-2 meters (39 -78 inches) per century incorrectly predicted.

The warming following the last phase of the Pleistocene ice age was accompanied by a rapid ice melt and increase in sea level over a period of about 7,000 years. Then, about 8,000 years ago the rate of sea level increase slowed dramatically and has been about the same since then. These facts are simply ignored by the prophets of doom. We often see statements in articles such as the following: "The rate of sea-level rise has tripled over the last decade, according to a recent study from the University of Miami, bringing with it more frequent coastal flooding."[29] The sea level there appears to be rising, but this effect is actually caused by the land sinking.

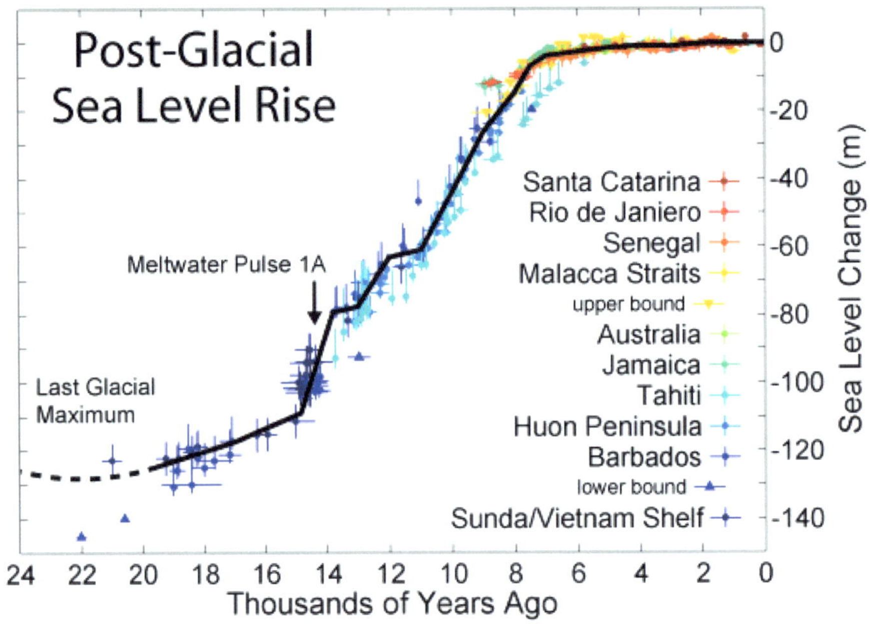

Because land can also be rising or sinking — a phenomenon called isostasy — sea level is tricky to measure. After the mile-thick Laurentide glacier covering Canada 20,000 years ago melted, that part of the North American Continent started rising, while the southern end of the continent bordering the Gulf of Mexico began to sink.[30] This includes the Miami area, which is why we've been told that sea level is increasing there.

Sea level measurements are made both by averaging tidal gauge data and by satellite altimetry. Dr. Mörner stated that the satellite data was tampered with to incorrectly show a rise in sea level.[31] Another system called GRACE – made up of gravitational-anomaly satellites - computes the mass of the ocean, from which changes in sea-level can be directly calculated.

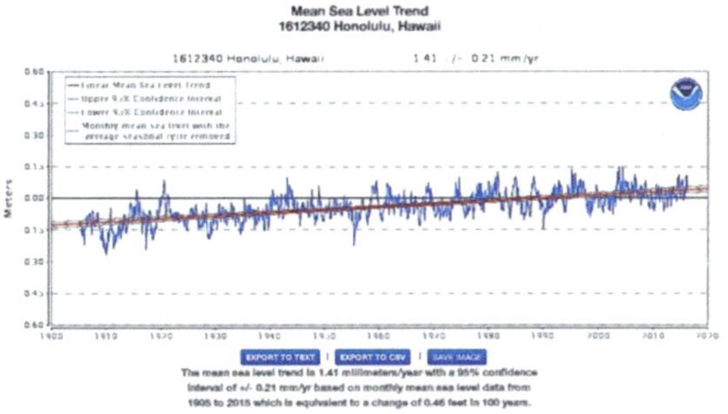

If the sea level was undergoing a significant increase, the line in this typical tidal gauge graph would curve upward. But instead it is a straight line. If anything, sea level rise is slowing down slightly.

Many media reports have falsely claimed that, due to the rising sea level, the Maldives Islands in the Indian Ocean are being submerged underwater and that its people will thus have to be relocated. However, Dr. Mörner stated, "Clear observational measurements in the field indicate that sea level is not rising in the Maldives, Bangladesh, Tuvalu, Vanuatu, or French Guyana.

The Beach at Tuvalu.

"The president of Tuvalu continues to claim that his islands are being flooded. However, tidal-gauge data provide clear indications of stability over the last 30 y ears.[32] In Vanuatu, the tidal gauge also indicates a stable sea level over the last 14 years."[33]

The graph of tidal gauge data in Honolulu is typical of what's been happening. "It would be difficult to convincingly argue that pre-1940 sea level rise was due to CO_2...So how can someone argue that sea level rise post-1940 was due to CO_2? (Sea level rise should be accelerating)"[34] Dr. Mörner stated that there's something "sick" in the IPCC's methods. "In science we discuss. We don't forbid or neglect...They are working with lobbyists. They blacklist those who disagree."[35]

On the last day of Barack Obama's presidency in January of 2017 NOAA issued a report that sea level could increase by eight feet by the end of this century. Then Mr. Obama promptly bought a $19-million beachfront estate in Massachusetts and a multimillion-dollar mansion on the beach in Hawaii, showing that he obviously wasn't concerned about any projected rise in sea level. Yet, despite the observational evidence — on which empirical science is based — the purveyors of doom continue to perpetuate this lie.

"Storms, floods, droughts and forest fires are much worse now due to man made climate change."

When, after the El Nino of 1998, the Earth's temperatures leveled out, the fear mongers seized on the idea of claiming that extreme weather events have been getting worse and more frequent. Since then The public has been continuously bombarded with media stories of such events.

Here again real science and statistics tell a different story. In fact, Bjorn Lomborg concluded that your risk of dying in a climate-related disaster has gone down 99% since 1920. This is because we have better systems in place to deal with these challenges. Dr. Roger Pielke, Jr., a mathematician, provided irrefutable statistical evidence clearly demonstrating that these claims are false. In sworn testimony before the US Senate Committee on Environment & Public Works, he laid out the following documented facts:

"It is misleading, and just plain incorrect," he said, "to claim that disasters associated with hurricanes, tornadoes, floods or droughts have increased on climate timescales either in the United States or globally. It is further incorrect to associate the increasing costs of disasters with the emission of greenhouse gases.

"Globally, weather-related losses [$] have not increased since 1990 as a proportion of GDP [they have actually decreased by about 25%], and insured catastrophe losses have not increased as a proportion of GDP since 1960.

"Hurricanes have not increased in the US in frequency, intensity or normalized damage since at least 1900. The same holds for tropical cyclones globally since at least 1970.

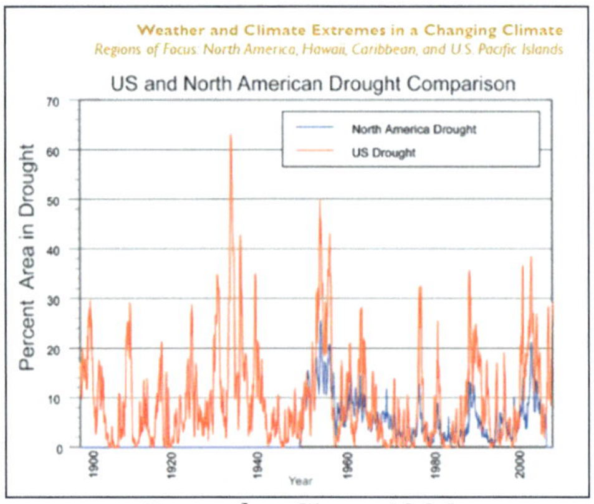

Droughts Since 1900

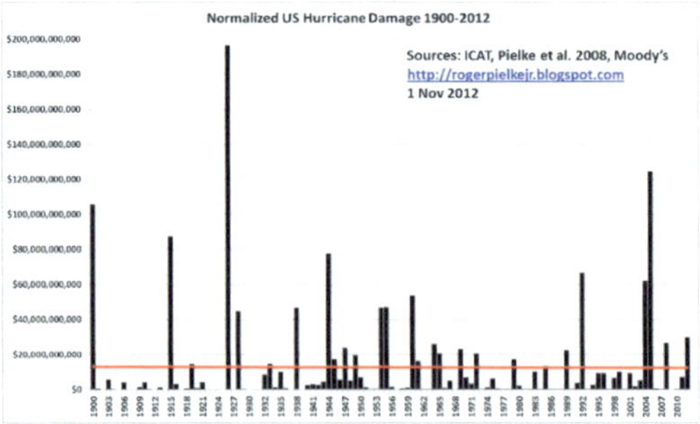

Record of Hurricanes

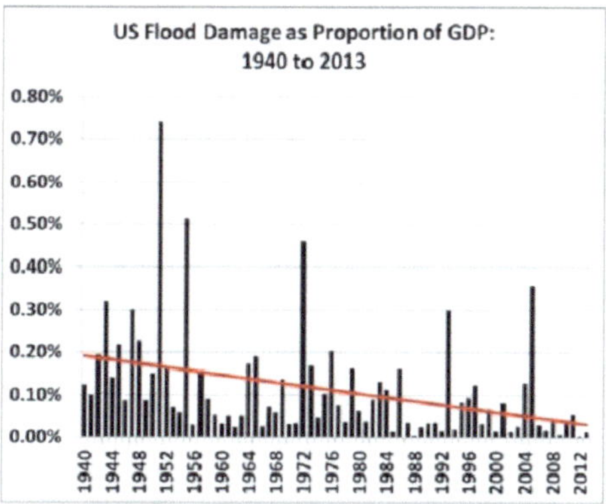

Floods Have Decreased.

"Floods have not increased in the US in frequency, intensity or normalized damage since 1950. Flood losses as a percentage of US GDP have dropped by about 75% since 1940.

"Tornadoes have not increased in frequency, intensity or normalized damage, and there is some evidence to suggest that they have actually declined. Drought has for the most part, become shorter, less frequent, and covers a smaller part of the US over the last century. Globally, there has been little change in drought over the last century."[36] These carefully compiled statistics were part of Dr. Pielke's sworn testimony before Congress. Because this information couldn't be disputed, global warming alarmists pulled out all stops to viciously attack his credentials, character and reputation. They claimed that he was "outside of the science community" and "on the fringe." The University of Colorado, where he is a professor, came to his defense.

Climate change alarmists frequently cite Hurricane Katrina and "superstorm" Sandy, as "proof of the effects of climate change." Katrina, which made landfall on Aug. 29, 2005, was a truly wicked hurricane, but it was far from the worst hurricane ever experienced. Much of the flooding damage was from levees that broke during the storm because they hadn't been properly maintained.[3] [7]Sandy was just a category-1 hurricane. But it did a great deal of damage due to a bizarre set of circumstances. First, as it moved up the east coast of the US in late October of 2012, it slammed into an extremely cold air mass moving south from Canada. This cold air mass added a good deal of energy to the system.

Sandy: a Category 1 Hurricane.

Sandy made landfall just south of Atlantic City, New Jersey, at the height of a spring tide, an unusually high tide that occurs when the sun and moon are lined up in such a way that they are both pulling in the same direction. "The storm surge — an additional 9 feet or more of water piled up against the coast by furious winds and crashing ocean waves — was riding on top of this extra high tide."[38] Strangely, adjacent areas that were hardly impacted at all by this storm experienced mysterious power failures or otherwise went offline.

As you can see from this list of the ten worst hurricanes in US history, the great majority occurred before the level of carbon dioxide in the atmosphere began to rise. The only recent ones were Hurricane Katrina in 2005 and Hurricane Ian in 2022.

1) 1900 Great Galveston Hurricane

2) 1928 Okeechobee Hurricane

3) 2005 Hurricane Katrina

4) 1893 Chenière Caminada Hurricane

5) 1893 Sea Islands Hurricane

6) 1881 Georgia/South Carolina Hurricane

7) 1957 Hurricane Audrey

8) 1935 Great Labor Day Hurricane

9) 1856 Last Island Hurricane

10) 1926 Great Miami Hurricane[39]

Hurricane Ian, which hit Florida, North Carolina and South Carolina in late September of 2022 was not the 4th strongest hurricane to hit Florida, but the 10th strongest, according to meteorologist, Dr. Neil Frank. "An analysis of major hurricane landfalls by decade shows a significant downward trend – not the upward trend predicted by climate change activists...Today's landfalls do not compare to those in the middle of the last century. Florida was hit by seven major hurricanes in the 1940s. Six major hurricanes slashed the east coast in the 1950s. But not a single major hurricane made landfall in the entire United States from 2005 to 2017."[40]

In 2022 National Public Radio announced, "Climate change poses an existential threat to the ski industry. A warmer climate means less snow and less now means a shorter season for snowboarders and skiers."[41]But at the end of February, 2023, there were back-to-back record-setting snowfalls in California that closed Yosemite National Park in the Sierra Nevada mountains. Scott Gediman, who has been a park ranger there for the past 27 years, stated that this was the most snow anyone there has ever seen. Again this demonstrates the absolute stupidity and cluelessness of the journalists who rely on false information from "experts".

sierranewsonline.com

Please don't take our word for this. Check out the facts for yourself.

1. https://www.worldwildlife.org/stories/polar-bear-population-decline-a-wake-up-call-for-climate-change-action
2. Amrutha Gayathri, "Polar Bear Population Higher than in 20th Century: Is Something Fishy about Extinction Fears? International Business Times, August 2, 2011
3. www.arcticwwf.org/wildlife/polar-bear/polar-bear-population/
4. IUCN/SSC Polar Bear Specialist Group, http://pbsg.npolar.no/en/index.html
5. Posted on 3/23/23
6. Ibid, p. 16
7. R.L. Holz, "How Many Scientists Does it Take to Write a Paper? Apparently Thousands," The Wall Street Journal, August 10, 2015
8. dso, op. cit., p. 18
9. Tol (2013), Montford (2013), Friends of Science (2014) and Legates et al (2015)
10. Idso, op. cit.
11. http://www.petitionproject.org
12. Hal Lewis, "Hal Lewis Resignation Letter From the American Physical Society," https:// www.heartland.org, Dec. 8, 2010
13. Hal Lewis, "Hal Lewis Resignation Letter From the American Physical Society," https:// www.heartland.org, Dec. 8, 2010
14. https://financialpost.com/opinion/open-climate-letter-to-un-secretary-general-current-scientific-knowledge-does-not-substantiate-ban-ki-moon-assertions-on-weather-and-climate-say-125-scientists
15. https://www.investors.com/politics/editorials/active-volcanoes-melting-west-antarctic-ice/
16. "NASA – The Mystery of Antarctic Cooling," YouTube video, Jan. 25, 2016
17. https://www.nasa.gov/feature/goddard/nasa-study-mass-gains-of-antarctic-ice-sheet-greater-than-losses
18. Laura Paddison, "Antarctic sea ice hit record lows again. Scientists wonder if it's 'the beginning of the end'", CNN.com, Feb. 2, 2023
19. Tony Heller, op. cit.
20. Professor Emeritus Don Easterbrook's eye-opening talk to the Washington State Senate committee on Climate Change, March 26, 2013
21. Ibid.

22. Seth Borenstein, " 'Arctic is screaming,' says scientists seeing new data", The HeraldBulletin.com, Dec. 11, 2007

23. Chaffin Mitchell, "NASA study reveals astonishing growth of one Earth's previously fastest-shrinking glaciers," AccuWeather.com

24. Tony Heller, "The 'Ice-Free Arctic' Big Lie", https://real climatescience.co m/2016/10/the-ice-free-arctic-big-lie/, Oct. 14, 2016

25. The Daily Mail, July 2016

26. Dr. Tim Ball, Human Caused Global Warming: The Biggest Deception in History, 2016, Tellwell Talent, p. 8

27. Tony Heller, RealClimateScience.com, March 17, 2023

28. https://www.noaa.gov/explainers/tracking-sea-level-rise-and-fall

29. Chris Bentley, "Facing rising sea levels and greater insurance risk, Southern Florida braces for relocations, new flood design standards, and more," http s://archpaper.com/2017/04/south-florida-rising-sea-levels, April 12, 2017

30. Dr. Tim Ball, "Sea Level Rise; A Major Non-Existent Threat Exploited by Alarmists and Politicians", WattsUpWithThat.com, July 21, 2018

31. Nils Axel Mörner, http://scienceandpublicpolicy.org/images/stories/ pa-pers/reprint/sea_level_ not_rising.pdf

32. Mörner, 2007ac, 2010b; Murphy, 2007

33. Ibid.

34. https://peg.atis.org/wp-content/uploads/sites/16/2018/08/Roy_Spencer.pd

35. Newman, op. cit.

36. Dr. Roger Pielke, Jr., Congressional Testimony, July 18, 2013

37. https://eng-resources.uncc.edu/failurecasestudies/dam-failure-cases/new-o rleans-hurricane-katrina-levee-failures/ (ASCE Review Panel 2007, p. v)

38. Malcolm J. Bowman, "Superstorm Sandy – How did it happen and are we prepared for the future?"https://seagrant.sunysb.edu/media/sandy 12/UUPI nsight-Sandy020113.pdf

39. Kim Bryan, "10 Worst Hurricanes in United States History," https://owlcatio n.com/stem/Worst-Hurricanes-United-States-History, Jan. 6, 2017

40. Neil L. Frank, "Did Global Warming Make Hurricane Ian Intensify Faster than Normal?", wattsupwiththat.com, Oct. 6, 2022

41. Kirk Siegler & Aaron Scott, "Can Skiing Survive Climate Change?" NPR, April 15, 2022.

Chapter Five

Natural Causes of Climate Change

> "One can be easily deceived by deferring to the authority of supposed experts." – Richard Feynman, Nobel Laureate

A graph of climate history plotting the amount of CO_2 with the average temperature shows that a rise or fall in temperature always occurs before the rise or fall in carbon dioxide levels. Proxy data (other than direct measurement by thermometers) taken from ice cores drilled in Antarctic and Greenland glaciers reveal the temperatures and percentages of various gases at different times in geologic history to around 900,000 years ago.

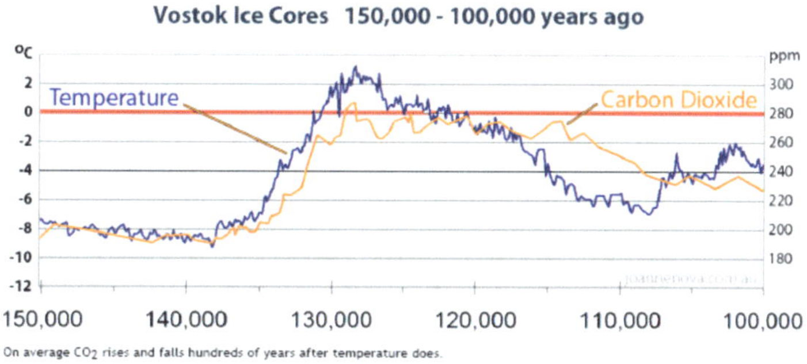

Temperature changes precede changes in carbon dioxide levels.

Data-gathering technology in ice core samples obtained before 1999 lacked the precision to determine this relationship between CO_2 and temperature. Improvements in this technology clearly showed that increases in temperature **always** precede increases in CO_2. This fact, <u>known by all climatologists by the year 2003</u>, demonstrates that an increase in CO_2 is the result, not the cause of increased temperature.

The average lag time between temperature increase and an increase in CO_2 is around 800 years, as shown on this graph. Seventy percent of the earth is covered by oceans. The 800-year lag time is caused by the fact that water, because of its high specific heat, cools and warms up much more slowly than land does. So it takes about 800 years to catch up to naturally-caused temperature changes on land. The oceans are warming now in response to the Medieval Warm Period ending around 1250 AD. The oceans release about 780 billion tons of CO_2 per year.

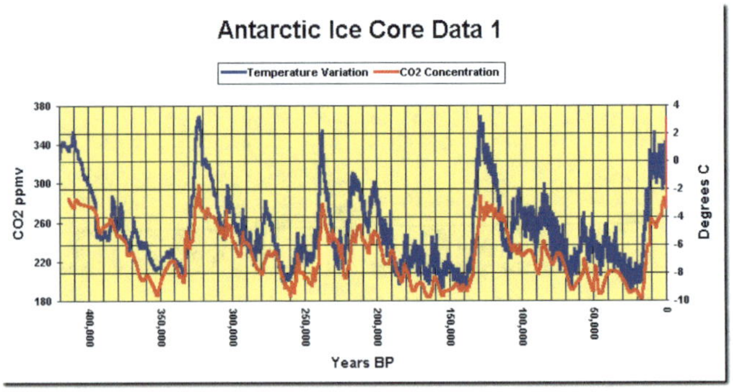

Temperatures precede changes in CO2

This is irrefutable proof that increases in carbon dioxide don't cause increases in temperature. It is just the opposite. It is the reason why careful measurements are absolutely crucial to any scientific investigation.

Ice core data from Greenland's glaciers also indicate that during most of the past 10,000 years Earth's temperatures were warmer than today. Ten thousand years ago the Arctic was ice free. In the past 500 years there have been about twenty periods of warming and cooling.

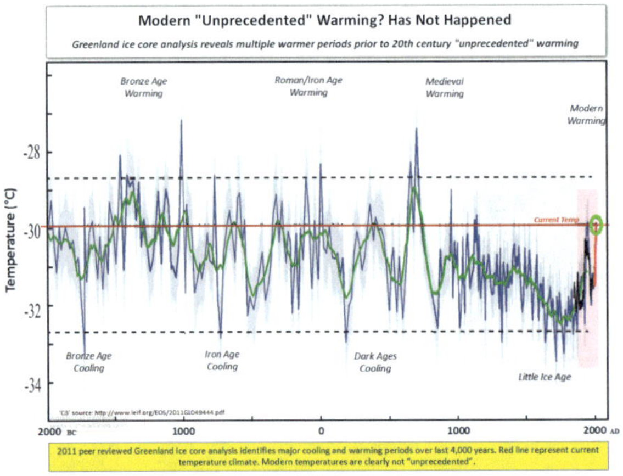

Temperature record from 2000 BC

During recorded history there was a warm period lasting from 250 BC to about 400 AD. This was appropriately named the Roman Warm Period. Some modern coastlines have been found two miles out from where they were during the Roman invasion of Britain in 43 AD, showing that at that time sea levels were significantly higher then than they are now.[1] Arctic temperatures were at least two degrees higher at that time than what we are seeing today. Amazingly, at that time in history the western Arctic ocean was covered in sea ice only about 50% of the year. Now it is ice covered for over ten months each year.[2]

During the ensuing period from 400 AD to 950 AD, known as the Dark Ages, it was colder. Then came the Medieval Warm Period, extending from about 950 AD to 1250 AD. Ice core samples show that although the average temperature was about two degrees warmer during the Medieval Warm Period, the carbon dioxide level was actually lower than it now is. This again demonstrates the lack of correlation between the percentage of carbon dioxide and Earth's average temperature. At that time a colony of Vikings from Scandinavia raised vegetables in a part of Greenland now covered by a glacier.

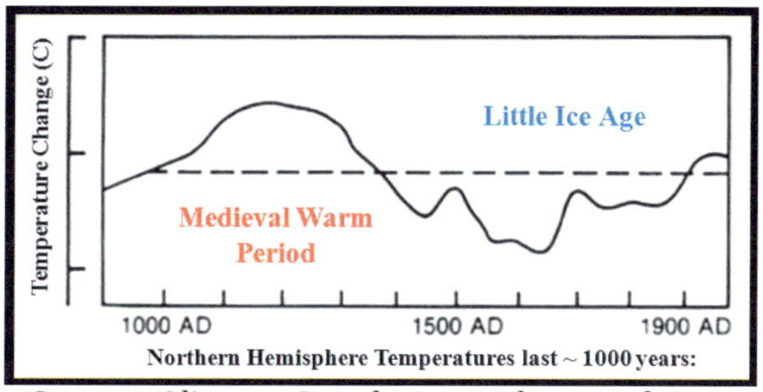

Correct Climate Graph: Last Thousand Years

Empirical data contradicts the doomsayers' claim that the Medieval Warm Period wasn't a worldwide phenomenon. Wikipedia falsely states that this phenomenon occurred only in the north Atlantic region. A thousand years ago the snow line in the Rocky Mountains was 1000 feet higher than it now is, and forests in what is now Canada were growing north of the location of the current tree line. A 5000-year-old white spruce was found 100 km north of the current tree line in Canada.[3] Permafrost prevents trees from growing north of that line. 1000-year-old tree stumps have also been found in ice caves under the Mendenhall glacier near Juneau, Alaska.[4] A forest was growing there 1000 years ago when it was warmer.

Then, after 1250 AD temperatures began to fall again, and by 1350 AD the Vikings' Greenland settlement had been abandoned. Shortened growing seasons and torrential rains in Northern Europe led to the "Great Famine" of 1315-1317 AD.[5] Temperatures continued to decline, and the very cold period known as the Little Ice Age lasted until the late 1700s. Ever since that cold spell ended, the earth has been in a gradual warming trend. Average temperatures have increased about 0.5°C. per century ever since, and scientists really can't explain why.

From the beginning of the 20th century until now there have been four periods of warming and cooling.

1. Between 1910 and 1940 average temperatures increased and were actually warmer than they are now. At that time, however, the level of CO_2 in the atmosphere was only 0.032% (320 ppm), whereas now it is above 400 ppm. Temperatures in the Arctic were higher during the 1930s than they are now and glaciers were melting.

2. In the 1960s and 1970s Earth's temperatures were decreasing at the same time that CO_2 levels were rising. Arctic ice rebounded during this period of cooling. How could this have happened if carbon dioxide was really the climate control knob?

3. Then, from 1978 to 1998 another period of warming occurred. CO_2 was also continuing to increase during that time period, causing some scientists to mistakenly conclude that this rise in carbon dioxide was causing that temperature increase. But they evidently failed to note that the sun was unusually active during that 20-year period of time.

4. Due to a super strong El Niño, a natural occurrence, 1998 was unusually warm, causing a temperature spike. In El Niño conditions, a band of warm ocean water develops in the central and eastern part of the Pacific ocean. El Niños, lasting for a year or so, cause Earth to warm slightly. This weather phenomenon occurs about once every 8-10 years.

Between 1998 and 2023 average temperatures remained fairly constant, perhaps cooling slightly, even though *about 30% of all human generated CO_2 was emitted during this time period.* In 2016 there was another El Niño, raising the average temperature for that year.

Of the above four time periods, #2 and #4 counter the human-caused global warming hypothesis. The decade of the 1930s, the warmest in the last century, occurred during period #1 when the carbon dioxide level was much lower than it now is. In the 1960s and 70s the earth was cooling while the CO_2 level was rising. 2013 was the coldest winter in 100 years in Europe. Five of the six snowiest winters in United States history occurred within the last decade.

For the first fifteen years of this century the IPCC, CRU and other agencies acknowledged that there had been no global warming since 1998. Nevertheless, to everyone's surprise, in 2015 a paper by Tom Karl of NOAA, apparently timed to coincide with the upcoming Paris conference, claimed that the temperature data showing this lack of warming never existed and that temperatures had been rising during that time. So Dr. Karl changed history.

Dr. John Bates, a respected NOAA climate scientist, came forward to reveal that this paper, which became known as the "Pausebuster Paper", was based on unverified and misleading data. Nevertheless, Karl's paper had a tremendous influence on policy makers from several nations who were trying to decide just how to commit their governments to dealing with what they were told was a crisis.[6] These people trusted the "experts", not realizing that other experts disagreed with them.

The period from October, 2018, to April, 2019, was the coldest and wettest 6-month period in the United States in over a century.[7]

Both RSS satellite and ARGO temperature measurement reveal that in the first 18 years of the 21st century temperatures remained fairly constant and may have even cooled a bit. By 1999 evidence surfaced that the amplification of the effects of CO_2 by water vapor originally predicted by climate alarmists was not happening. This fact was finally released in an obscure science journal in 2006 but was ignored by the climate alarmists.[8]

Many climatologists believe that we're living in an interglacial period — between two parts of the Pleistocene ice age — the next part of which could resume within as short a time as a thousand years from now.

You can now investigate the climate record for US reporting stations by accessing a free graphing program, RealClimateTools.com., by Tony Heller. Here you can access data from thousands of weather stations going back to the year 1900. You can look at the trend line for any of these stations to determine whether the temperatures in that community have been rising or decreasing. This is extremely interesting and informative. You can also observe the paths of tropical storms and hurricanes as well as their wind speeds for various years.

A team headed by Dr. Richard Lindzen of MIT conducted a study, published in 2009, showing that when the earth receives more heat from the sun, it simply radiates more heat back into space.[9]

Another climatologist, Roy Spencer, reported in 2007 that during periods of warming there are fewer high clouds to hold in heat.[10] This acts as a buffer that counteracts any feedback from increased water vapor that climatologists were predicting in the 1980s. Low clouds reflect some of the sun's radiation, cooling the earth.

Clouds are really the key to understanding climate. They can either hold in heat or block sunlight from reaching Earth's surface, cooling it. This depends on their height. High clouds generally slow down the dissipation of heat into space, causing warming.

On the other hand, low clouds block sunlight, which is then reflected back into space. This is called the albedo effect. The same thing happens when sunlight is reflected by ice and snow. If more low clouds than high clouds form as a result of extra humidity caused by CO_2, the earth is cooled. This is a *negative* feedback which offsets the slight warming effect of CO_2.

Beginning in the 1980s computer models also predicted that if CO_2 was causing warming, there would be a "hot spot" over the tropics, called a global warming "signature". The idea was that the resulting warm water vapor in the tropics would rise to a higher level of the troposphere, the lowest level of the atmosphere, about six miles up. For decades thousands of radiosonde balloons sent aloft in that area have failed to detect this hot spot. Many climatologists now feel certain that if that hot spot existed, it would have been detected by now.[11]

There are five different temperature datasets: two sets of satellite temperature records from the University of Alabama in Huntsville (UAH) and three land-based datasets: Hadcrut in England, NCDC and the NASA GISS temperature dataset. Anyone can download this data from the Internet, put it onto an Excel spreadsheet and run a trend line through the data.

Satellite data, which has been used since 1979, has proved to be a more accurate indicator of the earth's average temperature than surface weather stations for several reasons. First, thermometers used in ground-based weather stations are often located in cities close to air conditioner vents that spew warm air as well as in airports and other paved areas, where concrete heated by the sun raises the temperature. A century ago, most weather stations were located in grassy, unpaved, rural areas.[12] The influence of human activities such as traffic, concrete and air conditioner exhaust vents in cities is known as the urban heat island (UHI) effect. This can make a difference of up to several degrees in temperature measurement, skewing the results. This is because cities grew up around weather stations that had originally been in rural areas. A team of volunteers headed by meteorologist Anthony Watts checked US weather stations to spot those with such problems.

Also, surface weather stations can't measure Earth's average temperature accurately, because they're not evenly distributed around the globe. Most of them are located in the United States and Europe. There are far fewer weather stations in the southern hemisphere, the Arctic region and on oceans, which make up 70% of Earth's surface. So in computing Earth's average temperature, many temperatures in those regions are simply guessed at. Land based weather stations cover only about 25% of the earth. Thus, statements that we just had the "warmest year on record", when the average temperature has been recorded as *a few hundredths of a degree higher than the previous estimate* are meaningless. They simply indicate that temperatures were flat.

Temperature records have been kept for only about 140 years. Other methods of estimating temperature, such as tree rings and ice core samples, show the many temperature changes during Earth's climate history.[13] In contrast to land based thermometers, satellite temperature sensing equipment measures all of the earth's surface equally. The ARGO system has been used since 2003 to measure ocean temperatures.

Some warming in selected areas has been caused by changes in land use, especially the destruction of the rain forests and the growth of cities. The average temperatures of the earth's great cities, such as New York, have risen significantly in the past century. Yet the average temperatures of many smaller cities and towns have been flat or even lower during the same time period.

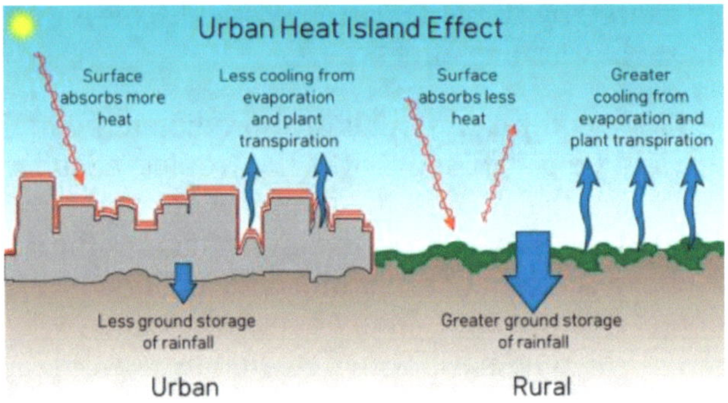

The United Nations commissioned the IPCC to study average temperature increases due ***only*** to the carbon dioxide emitted by human activities. They made no allowance for natural causes of climate change: cyclical changes in solar activity, changes in the earth's orbit caused by the gravitational pull of Jupiter, changes in the tilt of Earth's axis of rotation and the precession of the equinoxes. Just as a spinning top wobbles, the earth's tilt at a particular time in geologic history can range from 21 to 24 degrees. The tilt of the earth's axis is what causes seasons. At this time in geologic history the earth is closer to the sun in early January when the northern hemisphere is facing away from the sun. So winter cold in the northern hemisphere isn't as severe. But at other times in geologic history the opposite case occurs. Then the earth is farther from the sun during winter in the northern hemisphere. What did these people think caused the ice ages?

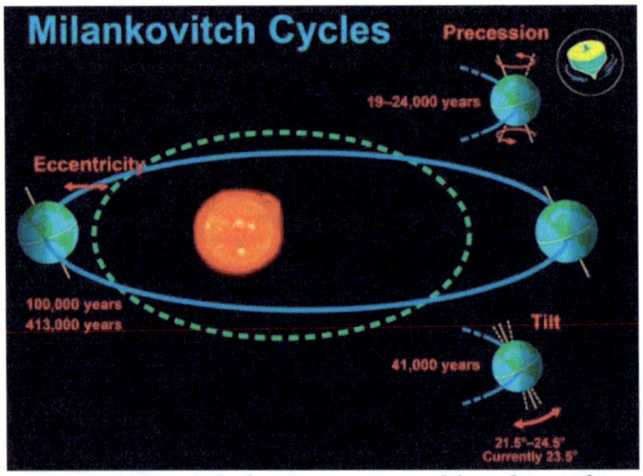

Precession and changes in Earth's orbit affect climates.

Since the 1990s Dr. Henrik Svensmark and his team have been studying the connection between the solar wind and cosmic rays. Sunspots, huge magnetic storms on the sun's surface, cause an increase in its magnetic field during the peaks in its 11-year sunspot cycles. When there are more of these magnetic storms on the sun, it releases more charged particles, called the "solar wind", into space. A tremendous number of these particles reach Earth. The solar wind, working with Earth's magnetic field, deflects cosmic rays heading toward Earth from outer space.

Cosmic rays are not electromagnetic rays but highly energized atomic-sized particles entering Earth's atmosphere at nearly the speed of light. These are believed to originate from huge explosions called supernovae, in which massive stars end their lives.

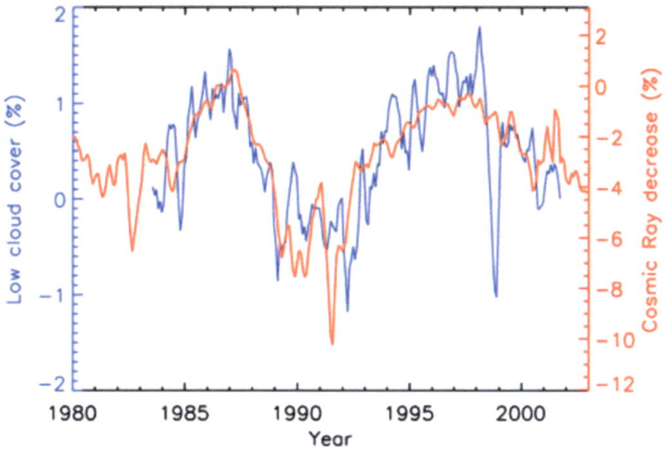

Correlation of cosmic rays and Earth's cloud cover.

Svensmark and his colleagues hypothesized that cosmic rays entering the atmosphere interact with organic molecules produced by trees to provide tiny particles called condensation nuclei around which water vapor condenses to form the water droplets that make up clouds. Without condensation nuclei clouds cannot form.

When there is less solar wind, during the low part of the sun's 11-year cycle when sunspots are at a minimum, more cosmic rays are thus able to enter the earth's atmosphere. During the 22-year period between 1983 and 2005, the correlation between cosmic rays and increased cloud cover was found to be 92%: extremely close. This has also been the case during the 400-year history of sunspot observation. When there are fewer sunspots — and thus more cosmic rays reaching Earth — there is also a higher percentage of low clouds, which reflect sunlight and cool the earth.[14]

CERN laboratory in Switzerland, which studies collisions between fast-moving atomic-sized particles, tested this theory and verified it. You will never hear this from the media. Another fact the major media haven't reported is that when average temperatures on Earth were cooler, as they were in the 1970's, other planets in the Solar System, such as Venus and Mars, also had lower average temperatures. This demonstrates that solar activity is a factor driving natural climate change.

Not all solar cycles are the same. In strong cycles there are many large sunspots, and in others hardly any. Solar cycles #22 and #23 were very active, causing the 20 years of warming near the end of the 20th century. At the end of solar cycle

#24, however, the sun was in a period of very low solar activity with no sunspots over a period of several days. The solar minimum between solar cycle #23 and #24 saw 260 spotless days in 2009 — the most days without sunspots in a single year since 1913.[15] During the Little Ice Age, a prolonged period of low sunspot activity known as the Maunder Minimum preceded an era so cold that between 1645 and 1715 the Thames River in London froze up in the winter, and people regularly ice skated on it.[16]

The sun goes through other cycles besides the 11-year cycle. Otto-Carl Weiss and his colleagues examined European temperature data for the past two hundred fifty years using Fourier Transform spectral analysis. Their measurements indicated that, rather than a constant increase in temperature since the beginning of the Industrial Revolution, the climate had warmed and cooled periodically in ~200-year De Vries solar cycles. The team also concluded that climate change is natural with no influence from carbon dioxide.[17]

Another theory of the natural cause of climate change was introduced by Dr. Don Easterbrook, who conducted an extensive investigation of all warm/cold periods, not just full-scale Ice Ages and interglacial warm periods but also short-term warming and cooling periods.

He states: "For each warm/cold period, I looked at temperature, sunspot numbers (SSN), total solar irradiance (TSI), and radiocarbon (14C) and beryllium-10 production rates as indicators of cosmic ray intensity. The data are truly remarkable. Every cold period, regardless of the magnitude of the warm or cooling, was characterized by low sunspot numbers, low TSI, and high production rates of 14C and 10Be, indicating high cosmic ray intensity. Every warm period was characterized by high sunspot numbers, high TSI, and low production rates of 14C and 10Be, indicating low cosmic ray intensity."

Dr. Easterbrook's study supports the conclusions of the Svensmark Clouds study, but does not depend on them. It stands alone on its own merits. The results of this study were summarized in the book "*Solar Magnet Cause of Climate Changes and the Origin of the Ice Ages*".[18]

Patterns of ocean currents are a large factor in minor warming and cooling of the climate. The Pacific Decadal Oscillation (PDO) is one such current. It reverses every 20 to 30 years.[19] When warmer water from the south Pacific moves through the Bering Strait into the Arctic Ocean, the Arctic is warmed and ice melts. The Pacific Decadal Oscillation, largely influencing United States climate, switched to this pattern in 1910, after which warming occurred, and again in 1977, after which the warming pattern repeated. It is amazing how these two dates line up with the beginning of each warming period occurring in the 20th-century.[20]

---

1. http://notrickszone.com/2018/04/26/arctic-temps-2-6c-warmer-than-today-with-4-5-fewer-months-of-sea-ice-coverage-2000-years-ago/#sthash.oUxXZuIP.dpbs
2. .L. McKay, et al , Holocene fluctuations in Arctic sea-ice cover: dinocyst-based reconstructions for the eastern Chukchi Sea1-2
3. Robert Zubrin, Merchants of Despair, New York: Encounter Books, 2012, p. 227

4. Ball, op. cit.

5. Larry Bell, "The New York Times' Global Warming Hysteria Ignores 17 Years Of Flat Global Temperatures," Forbes Magazine, Aug. 21,2013

6. Anthony Watts, "Bombshell – NOAA Whistleblower says Karl et al. "Pause-buster" Paper Was Hyped, Broke Procedures," WattsUpWithThat.com, Feb. 4, 2017

7. Tony Heller, YouTube video, "Coldest October-April In The US In Over A Century," May 2, 2019

8. Topher Field, "50 to 1David Evans Interview," YouTube, Sept. 1, 2013

9. Knappenberger, https://www.masterresource.org/.../climate-sensitivity-est imates-heading-down-way-down

10. Joannenova.com.au/2009/12/the-one-flaw-that-wipes-out-the-crisis/

11. David Evans, The Australian, July 18, 2008

12. David Evans, "Is the Western Climate Establishment Corrupt?", Nov. 8, 2010, http://science and publicpolicy.org/images/ stories/papers/originals/western_climate_establishment corrupt.pdf

13. Ball, op. cit., p. 8

14. Larry Vardiman, A New Theory of Climate Change, 11/1/08 https://www.ic r.org,

15. http://www.climatedepot.com/2018/05/23/are-we-headed-towards-another -deep-solar-minimum/

16. Mark Prigg, "Is a Mini Ice Age On the Way?", DailyMail.uk, July 10, 2015

17. Lecture by Prof. Carl-Otto Weiss, Advisor to the European Institute for Climate and Energy, "International Conference of the Schiller Institute", Paris, France; June 14, 2015

18. Dr. Don Easterbrook, from abstract, July 9, 2019

19. https://sealevel.jpl.nasa.gov/data/el-nino-la-nina-watch-and-pdo/pacific-de cadal-oscillation-pdo/

20. Dr. Fred Goldberg lecture March 15, 2016 (YouTube video)

Chapter Six

The Politics of Climate Change

> "We know that no one ever seizes power with the intention of relinquishing it. Power is not a means. It is an end." — George Orwell, *1984*

A Canadian oilman who made a huge fortune in the US, Maurice Strong played the most important role in establishing the Intergovernmental Panel on Climate Change (IPCC) in 1988. A friend of the oil magnate David Rockefeller, Strong was appointed chairman of the 1992 Earth Summit United Nations Conference, at which he advocated population reduction and the lowering of living standards in the interest of "saving the environment." His radical viewpoints were revealed years later when he asked, **"Isn't the only hope for the planet that the industrialized civilizations collapse? Isn't it our responsibility to bring that about?"**[1]

Strong helped found the Club of Rome in 1968. The members of this organization believed that the earth's resources were being used up too fast, especially by the industrialized nations. Many of their beliefs stemmed from the mistaken idea first advanced by Thomas Malthus in his 1798 *Essay on the Principle of Population* that the world's population would outgrow the food supply. Malthus' conclusions were disproved by what actually happened. As the world's population increased, the percentage of people who were thriving also increased.

Maurice Strong

Malthus' deluded philosophy contributed to the ideas of Charles Darwin[2] and the development of the eugenics movement in the early 1900s. This movement, based on warped thinking, advocated the selective breeding of the human population – by methods such as the sterilization of those deemed inferior – to improve its genetic composition. But who would decide who was superior and who was inferior? Obviously it would be those in power. This led to the unfortunate episode of the Holocaust carried out by the Nazi's in World War II.

In his book, *The Population Bomb*, published in 1968, Paul Ehrlich revived Malthus' idea. In this book Ehrlich predicted, "The battle to feed humanity is over. In the 1970s, the world will undergo famines. Hundreds of millions of people are going to starve to death in spite of any crash programs embarked upon now. Population control is the only answer." None of Ehrlich's forecasts materialized. Yet there are still many people, even some billionaires[3], who believe the world's population should be as low as 500 million – not the current 7.5 billion. They're afraid that if Third-World people get the energy they need to raise their living standards, the earth's population will rise further.

Paul Erlich

Actually the opposite is true. History has shown *in every instance* that when a population's living standards increase, the birthrate decreases.[4] First the death rate comes down; this is followed in a few years by a decrease in the birthrate.[5] This principle is known as the Demographic Transition, as shown in this graph.

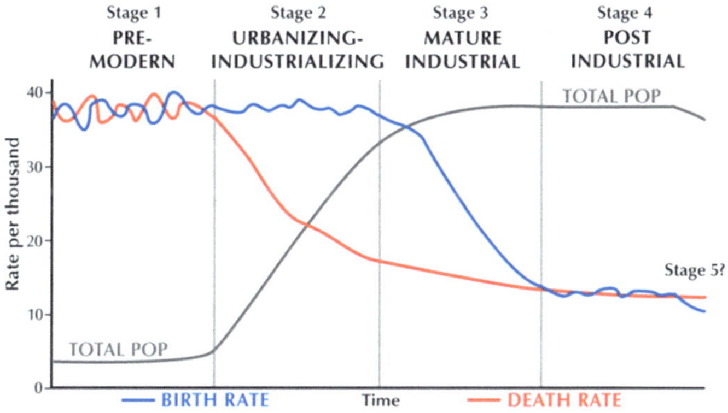

History has repeatedly demonstrated that industrialized civilizations have lower birth and death rates. In asserting that the world's rapid population growth was unsustainable, Ehrlich proposed radical population control measures, including sterilization. In 1977 he and John Holdren, who later became Obama's science czar, wrote a book entitled, *Ecoscience: Population, Resources, Environment.* This book called for forced abortions, contraception drugs in water supplies, seizing babies from single and teen mothers, forced sterilization of those they deemed unfit and an international police force.[6] This concept of eugenics, which inspired the Holocaust perpetrated by the Nazis, has been around for a long time.

The Club of Rome's frightening goal and strategy based on these mistaken ideas was revealed in its book, *The First Global Revolution* (1991), which asserts, "In searching for a new enemy to unite us, we came up with the idea that pollution, the threat of global warming, water shortages, famine and the like would fit the bill...All these dangers are caused by human intervention...The real enemy, then, is humanity itself."[7] Here they had hit on something for which to make people feel guilty. A well known strategy used by governments for grabbing power that has been known and used for a long time is:

PROBLEM → REACTION → SOLUTION

Those in power create a problem or crisis that is real or imagined. They sell this "problem" to the people using thousands of reports and articles — cleverly contrived propaganda. The public reacts by begging for a solution, which is then provided handily by the people in power. The rulers say, "We're here to solve the crisis we created." Because global warming is a worldwide "problem", the "solution" is too difficult for nation states by themselves to handle. Thus, the solution to this particular "problem" is nothing short of totalitarianism on a global scale.

Just as in the book, *1984*, a common enemy and a common cause around which the entire planet could rally had been found. Environmentalism, an important and noble cause, has been hijacked by the ruling elite who have no interest in the environment, as demonstrated by their huge mansions and private jets in which they regularly fly to exotic places for climate conferences. After the first Earth Summit, Strong was appointed Assistant Secretary General of the United Nations. In 1992 he organized the UN Framework Convention on Climate Change (UNFC-CC) at the Rio Earth Summit. It was at this conference that the globalist-inspired UN "Agenda 21" (now called "Agenda 2030") was also established.[8]It's no accident that the UNFCCC and the "Agenda 21" plan were given birth at the same time. The UNFCCC and Intergovernmental Panel on Climate Change (IPCC) would provide the excuse for the establishment of "Agenda 21". According to the investigative journalist, Elaine Dewar, "Strong was using the UN as a platform to sell a global environment crisis and the global governance agenda."[9]

Strong also set up the United Nations Environmental Program (UNEP) and the World Meteorological Organization (WMO) to create the science he needed to advance the Club of Rome's political agenda. The Intergovernmental Panel on Climate Change (IPCC) would provide policy recommendations for politicians. Climate scientists were appointed to this agency to research the subject and write reports which would lend the scientific authority for the policy reports to be handed to individual governments.[10]

The UNFCCC literally controlled the outcome of this research by narrowing its focus. In their research the IPCC scientists were permitted to consider only the IPCC's definition of the term "climate change", which they defined as: "a change of climate *which is attributed directly or indirectly to human activity* that alters the composition of the global atmosphere and which is, in addition to natural climate variability, observed over considerable time periods."

These scientists were not allowed to consider any other possibilities. The IPCC would write its summary report before the scientific report had even been submitted. They would take that report to the scientists and tell them, "Make sure your report agrees with this." In their first report, written in 1995, the IPCC scientists truthfully acknowledged that, although the climate was changing, there was no direct evidence that any recent changes had been caused by mankind.

Despite this conclusion by the IPCC scientists, the final policy report summary composed by Dr. Ben Santer (of Climategate fame: see Chapter 2) reads: "There is evidence of an emerging pattern of climate response to forcing by greenhouse gases and sulfate aerosols...from the geographical, seasonal and vertical patterns of temperature change...These results point toward a human influence on global climate." So Santer disregarded the conclusions that the IPCC scientists had reached and wrote his own report.

Although most IPCC scientists are competent and truthful, few have had any influence over final IPCC conclusions. Their work is cherry picked line by line by political appointees, many of whom are not scientists but activists from environmental groups such as the World Wildlife Fund. These "Summaries for Policymakers" always end up saying what the IPCC wanted them to say all along and are then passed on to the media and the public. This caused Frederick Seitz, a world-famous physicist and former president of the US National Academy of Sciences, the American Physical Society and Rockefeller University, to write, "I have never witnessed a more disturbing corruption of the peer review process than events that led to this IPCC report."[11]

Other top scientists who had formerly supported the IPCC have protested its unscientific methodology. Professor Fritz Vahrenholt, a German environmentalist, also criticized the IPCC for its shady practices, incompetence and dishonesty in its gross exaggeration of the influence of human CO2 emissions on the climate. He wrote a book entitled, *The Cold Sun: Why the Climate Disaster Won't Happen.*

In 1993 US Senator Tim Wirth (D-CO), who was on board with all of this, revealed the true motivation for this deception. "We've got to ride the global warming issue," he said. "Even if the theory of global warming is wrong, we will be doing the right thing." Wirth had arranged for NASA's James Hansen to deliver testimony before Congress about global warming on June 23, 1988, a day they expected to be one of the hottest of the year. Later he candidly admitted on an ABC "*Frontline*" broadcast, "We went in the night before and opened all the windows, I will admit, right, so that the air conditioning wasn't working inside the room...it was really hot."[12]

Sometimes the perpetrators of a deception put the truth right out in plain sight. In 1998 Christine Stewart, Canadian Minister of the Environment, commented, "No matter if the science is all phony, there are collateral benefits...Climate change [provides] the greatest chance to bring about justice and equality in the world."[13]We have to ask why someone in her position would make a comment like that.

Despite this promise of "justice and equality," the fulfillment of the IPCC goals by means of the "Green New Deal" or a similar plan would actually have the opposite effect. The IPCC's stated aim is to redistribute wealth from the rich developed nations to the poor third-world nations. Looking more closely, this plan would really transfer funds from the poor of the rich nations to the corrupt rulers of the third-world nations, many of whom personally profit at the expense of their citizens.

In introducing *Agenda 21,* Maurice Strong boldly declared, "Current lifestyles and consumption patterns of the affluent middle class – involving meat intake, use of fossil fuels, appliances, air-conditioning, and suburban housing – are not sustainable."

In 1992 *Agenda 21* slipped into the United States under the radar. Since it was supposedly a "voluntary" plan, the people were never consulted or even informed of this trickery. In 1993 the Clinton Administration wasted no time in establishing the "President's Council for Sustainable Development" to begin implementing some of its goals.[14] Congresswoman Nancy Pelosi introduced a resolution of support for the plan into Congress, calling it a "comprehensive blueprint for the reorganization of human society."[15] *Agenda 21* has now morphed into the Green New Deal. Global warming is the excuse.

Although "social justice" is their byword for grabbing the moral high ground, the elites have denied loans to African countries such as Nigeria to install desperately-needed coal fired power plants. If they use coal to produce electricity, Nigeria's thirty million people will lose the aid they receive from the European Union. These nations must promise to install only "alternative" energy that is impractical and 2-5 times as expensive.

As a result, the over 1.2 billion people who still don't have electricity must pay a high price for it. Only the rich can afford that luxury. The poor are blocked from using cheap, abundant energy to pull themselves out of poverty. The price of food has skyrocketed, partly because some food crops, such as corn and sugar cane, have been diverted to the production of ethanol and other biofuels. (See Chapter 8) This misappropriation of agricultural land caused food riots in Egypt, directly causing "Arab Spring." Is this social justice?

The well-placed elite have also instituted carbon trading, a practice yielding billions of dollars. In this scheme, also called emissions trading, or "cap and trade", businesses are forced by governments to trade their "pollution rights" through a complicated system of regulations. The companies producing less CO_2 can sell their unused "pollution rights" to companies that produce more carbon dioxide. This, of course, provides a financial incentive to produce less CO_2. But it is the consumers and taxpayers who ultimately pay the bill for this in the form of higher prices and taxes. During the last twenty years Al Gore, the main spokesman for human-caused climate change, increased his wealth from $2 million to about $300 million.

By the year 2008 $126 billion had already been paid out in these cap-and-trade schemes. This practice alone is expected to generate 2 to 10 trillion dollars for those set to benefit. Just one cap and trade auction held in California in February, 2022, generated about $975 million in revenue.[16] The cap and trade scheme was engineered by the bankers of such firms as J P Morgan Stanley, Barclays, Citigroup and Credit Suisse, who, in brokering these deals, are siphoning money away from hard working consumers and taxpayers. Casey Mulligen, reporting in the *New York Times*, stated that in the US "Some estimates suggest that the amount at stake for the taxpayers is [annually] over $3,000 per person."[17]

So the hypothesis of human-caused climate change, which has never been scientifically verified and cannot be disproved, has reaped huge profits for bankers, solar and wind corporations and companies set up to benefit from cap and trade. It has also provided careers for a new generation of scientists with a vested interest in this "problem". It has become obvious that the climate change agenda is not about climate change. It is about total control of the people.

In 1966 Carroll Quigley wrote in his book, *Tragedy and Hope*,

> "The global elites' aim is nothing less than to create a world system of financial control in private hands able to dominate the political system of each country and the economy of the world as a whole.[18]

The great climate change lie is moving the world one step closer to that reality.

---

1. WikiQuote.com/wiki/Maurice_Strong
2. The complete name of his famous book is The Origin of the Species by Means of Natural Selection, or the Preservation of Favoured Races in the Struggle for Life.
3. such as Ted Turner and Bill Gates, who gave a Ted talk calling for a drastic reduction in the world's population.
4. Ball, op. cit., p. 95
5. Steven Mosher interview, Population Research Institute, "Crossroads",Epoch Times 6/30/22
6. Dr. Tim Ball, Human Caused Global Warming: The Biggest Deception in History, Tellwell Talent, 2016, p. 85.
7. Alexander King and Bertrand Schneider, The First Global Revolution, 1991, Pantheon Books
8. Alexander King and Bertrand Schneider, The First Global Revolution, 1991, Pantheon Books
9. Ball, op. cit., p.102
10. Ibid.
11. Ball, op. cit., p.102
12. Steven Goddard, "Scientists Explain Why Hansen/Wirth Sabotaged The Air Conditioning In Congress In June, 1988," https://stevengod-dard. wordpress.com/2011/04/12/scientists-explain-why-hansenwirth-sabo taged-the-air-conditioning-in-congress-in-june-1988/, April 12, 2011
13. Zubrin, op. cit.
14. DeWeese, op. cit.
15. Ibid.
16. https://lao.ca.gov/Publications/Report/4591
17. https://economix.blogs.nytimes.com/2009/05/27/how-much-will-cap-and-t rade-cost/
18. Carroll Quigley, Tragedy and Hope: A History of the World in Our Time, New York: The Macmillion Co., 1966

Chapter Seven

Climate Change Propaganda

> "We will know our disinformation program is complete when everything the American public believes is wrong." — William Casey, Former CIA Director

"The impacts of climate change are already being felt in communities across the country."[1] This seemingly innocent statement raises a red flag. We need only examine the documented evidence to realize what a false claim this is. Despite what you may have heard on TV or read in articles, Satellite temperature records indicate that there has been no appreciable warming in the past 24 years.

Statistics show that the incidence of serious hurricanes, wildfires, tornadoes, drought, and floods is about the same or even less than before the amount of the trace gas carbon dioxide increased. (See Chapter 4.) The rate at which the sea level is rising — verified by NOAA — is about the same as it has been for the last 1000 years: seven inches (not twenty feet!) per century. There is no evidence that the pH of the ocean has been reduced. The ocean is not acidic but alkaline.

Al Gore, "An Inconvenient Truth"

Hurricane Katrina, a deadly hurricane that made landfall in 2005, was featured heavily in Al Gore's documentary, "An Inconvenient Truth", in 2006, just a year later. This hurricane was still fresh in everyone's memory, giving the film a tremendous impact. However, looking at the entire record, we know that the incidence of major hurricanes has actually decreased in recent years.

In 2007 Gore received the Nobel Peace Prize for this documentary at about the same time that a British judge, Michael Burton, listed nine significant errors in the "one-sided" production and ruled that it could be shown in British schools only if accompanied by teacher guidance explaining these errors to their students. Among other errors, Gore claimed that melting in West Antarctica or Greenland would cause a twenty-foot sea level increase in the near future.[2]

Nowhere is this outrageous propaganda more evident than in our schools. Just as the Nazis brainwashed the German children, the global warming movement, which hijacked environmentalism, has unfairly taken advantage of captive audiences in our schools to influence their young minds.

For example, on Friday, March 15, 2019, school children all over the world took the day off from school to participate in a demonstration to "save the world from climate change." The headmasters of schools in Britain suspended classes even though their students, were supposedly "on strike". It was obvious that this affair, a PR stunt arranged to garner more media attention, was organized by the adults.

This manipulation of young children's minds takes advantage of the fact that children are impressionable and lack the education and experience to discern truth from fiction. The strategy of the organizers is akin to that used by the Hitler Youth leaders in Nazi Germany.

A nightmarish example of this type of manipulation is David Suzuki's statement: "The North Pole, once a wintry wonderland, is no longer safe for Santa's Workshop. Climate change is melting the snow and ice, and the rising water is getting too close for comfort. Santa must relocate — fast — to make sure that all the nice boys and girls still have a Happy Holiday."[3] Dr. Patrick Moore calls this shameful exploitation of children to advance one's agenda "child abuse". He recently told the climate alarmists in a tweet:

> "You don't seem to get it, that this is a fascist tactic. You are using innocent children for your own ends. And this is child abuse of a psychological nature. It's normally called brainwashing, and you are guilty of participating in it. Mao would be proud...Y ou are clearly projecting your own psychosis on the children. Self-loathing is a heavy burden. Why don't you express it through adults rather than abusing innocents?"[4]

Propaganda is a persuasive mass communication technique designed to advance the interests and agenda of a particular group. It bypasses logic by using faulty reasoning – logical fallacies – and emotional appeal. We have a great responsibility to teach children, as well as adults, critical thinking skills to help them recognize the signs that they are being exploited.

Some logical fallacies (errors in logic) constantly used in man-made global warming propaganda are:

1. "ad hominem" (against the man) attacks, which either directly call names, such as "climate denier" or indirectly disparage those who disagree by assaulting their credentials or character – eg. "on the fringe".

2. “straw man” - distort your opposition’s argument and then knock it down. A “climate change denier” is someone who denies that the climate is changing. No skeptic ever said that.

3. “band wagon” – states that there is a consensus of the majority promoting the agenda, and that anyone who disagrees doesn’t belong. This works because of the basic human need to be on the “winning team”.

4. “appeal to authority” - We are constantly told to “trust the experts” without evidence. The telltale sign is the lack of information. We are never told that there are other experts who disagree with the “experts”. We are starting to learn from the nonsense that many (not all!) people with doctoral degrees are spouting, such as the ability to change one's gender ("trans"), that they lack common sense and the ability to think critically. Anyone who has passed Biology 101 knows that in every one of the ***trillions*** of human body cells, each male cell contains an XY chromosome and each female cell contains a XX chromosome. Despite the ingestion of hormones and multiple operations removing sex organs, it is utterly impossible for anyone to change to the opposite gender.

5. “moving the goalpost” - When nature failed to cooperate with the global warming models and Earth stopped warming after 1998, the global warming alarmists simply changed their slogan from “global warming” to “climate change” and even “climate disruption” to falsely imply that this “crisis” was causing storms, droughts, etc. Another way they’ve moved the goalpost is to keep advancing their “tipping points” to the future after their current tipping point passes without incident.

6. “appeal to sympathy” - blames skeptics for the plight of the polar bear and the victims of storms, droughts, floods and even forest fires, which, of course, are no worse than they’ve always been.

7. "guilt by association" These experts in psychology have tried to create a *false association* of climate skeptics with folks in groups like the Flat Earth Society, who believe the earth is flat. Of course, no climate skeptic believes such a moronic idea, but the constant repetition of their points sticks in the public consciousness after a while. This is truly unfortunate.

Other signs of this insidious propaganda are:

1. Weather events that are normal are presented as abnormal.[5] The climate events we’ve experienced in the last century are all within the parameters of natural climate change. The statistical record plainly shows that despite what the alarmists say, there has been ***no*** increase in droughts, floods, heat waves, wildfires, hurricanes or tornadoes.

One example is this headline: “The 2017 Hurricane Season Really Is More Intense Than Normal.” The key phrase here is “more intense than normal”.

2. Definitions are changed. This is a logical fallacy known as *equivocation*. For example, in 2004 climate alarmists, realizing that the warming had stopped, changed their mantra “global warming” to “climate change”.

This stroke of genius provided a greater range of possible things that could go wrong, making the "problem" more threatening. Now you see headlines claiming that it is either the warmest, coldest, wettest, driest, calmest, and windiest on record.

3. They play on our emotions. - fear, guilt and pity - rather than reason and logic. They tell us that we should feel guilty for what we are doing to our children and grandchildren. We should take pity on the poor polar bears.

4. They take advantage of the public's lack of science knowledge. Climate is a complicated subject. Even some scientists who specialize in other disciplines, such as biology, may lack the confidence needed to do their own research and make an informed decision. So they trust the "experts". This enables the climate alarmists to make up fake terms such as "polar vortex" and unscientific statements, such as "The science is settled." Real science is never settled.

5. They keep ramping up the threat to make it more and more ominous.

Here are some of the latest threat predictions:

"Global sea levels are projected to rise between 2 and 7 feet by 2050, and possibly more. By the end of the century, sea levels could rise as much as 20 feet. In this scenario, the United States alone would lose nearly 50,000 square miles, which today are home to 23.4 million people."[6]

"In 2020, scientists discovered a new source of greenhouse gas emissions off the coast of East Siberia, as warmer waters cause the seafloor to release ice-encased *methane, a potent source of global warming.* According to the U.S. Geological Survey, the so-called Arctic hydrate destabilization is one of the most serious scenarios for abrupt climate change that has implications far beyond the melting ice of the polar north."[7] But methane in the atmosphere is only 2 parts per million, with a negligible effect on the climate. (See Chapter 3) So this argument is preposterous.

A good example of this type of propaganda is the following editorial by Mark Hertsgaard: "Climate Denialism Is Literally Killing Us. The victims of Hurricane Harvey have a murderer—and it's not the storm."

> "The horrors hurled at Houston... in late August were heartbreaking — but also infuriating. How many times must we see this disaster movie—titled Hurricane Harvey in 2017, Hurricane Sandy in 2012, and Hurricane Katrina in 2005, along with many lesser — known foreign releases — before we intervene and change the ending? And how long before we hold the ultimate authors of such climate catastrophes accountable for the miseries they inflict?

> "The tragedy of Harvey starts with the suffering of innocents like Jordyn Grace, the 3-year-old who survived the flood by clinging to the body of her drowned mother, who had prayed with her last breaths. At least 60 people died in Texas because of the storm, over 1 million people were displaced, and who knows how many survived but lost everything? "What makes this so infuriating is that it shouldn't be happening...And how long before we hold the ultimate authors of such climate catastrophes accountable for the miseries they inflict?"[8]

The rant goes on and on. To whip people into an emotional frenzy, Hertzgaard blames this storm on "climate denialism", which itself is a misnomer. No one denies that there is a climate or that the climate is changing.

A category 4 storm, Harvey, in 2017, was the country's first major (category-3 or higher) hurricane to make landfall on the US mainland since hurricane Wilma hit Florida in 2005. This ended a twelve-year period, the longest in US history, with no major hurricanes. Hertsgaard either didn't do his homework or tried to push lies onto the American people in a deliberate attempt to instigate fear and misdirected blame.

This map of the United States appeared in the November, 2013 issue of National Geographic magazine. The projected sea level rise used in this map is based on a sea level increase of 262 feet that was predicted in 2000 by the US Geological Survey (USGS). But such a sea level increase could occur only if — in addition to the Greenland glaciers and Arctic sea ice — *all* of Antarctica's glaciers melted. If that were to happen, the earth's sea level would rise ~73 meters (about 240 feet).[9]

The periods of very slight warming within natural limits that have occurred in the last few hundred years have provided no evidence that anything like this could ever happen for a long time: millions of years. As a matter of fact, some climatologists believe that we are living in an interglacial period, between two parts of the Pleistocene ice age. So this article in National Geographic magazine is a figment of the imagination and a supreme example of propaganda. The editors of National Geographic should be ashamed that they ever published such an article.

Dr. Patrick Moore summarized what has been happening.

> "Fear has been used all through history to gain control of people's minds and wallets and all else, and the 'climate catastrophe' is strictly a fear campaign — well, fear and guilt — you're afraid you're killing your children because you're driving them in your SUV and emitting carbon dioxide into the atmosphere and you feel guilty for doing that. There's no stronger motivation than those two."[10]

In George Orwell's classic, 1984, a totalitarian government controlled its people with terror and propaganda lies and even by rewriting history. Some of the slogans used by the Party were: "War is peace. Freedom is slavery. Ignorance is strength." The people who had been carefully taught not to question anything went along with this party line. Now in 2023 comes the frightening realization that the theme of this story is coming true. The public is now being told that the frigid winters we've experienced in the past several years indicate that Earth is warming.

This is a good example of climate change double talk. A recent CNN report claimed that global warming doesn't always mean warming. "The term 'global warming' confuses people, because it triggers thoughts about warmth, and it sort of lends itself to misinterpretation when it also impacts the cold," declared Mike Hulme of the University of Cambridge. The message from this is "Don't trust your senses or your instruments."[11]

When skeptics point to several straight years of severely cold winters, they are reminded that weather isn't climate. "Not only are severe snowstorms possible in a warming climate, they may even be more likely...These storms draw much of their intensity from the extreme contrast between cold air over land and warmer, wetter air from over the ocean," asserted a NOAA spokesperson. So we are being told to suspend scientific observation.[12] But a heat wave that is entirely within natural limits is used for an article warning us that climate change is destroying the planet. So whatever happens: whether it's searingly hot or frigidly cold, whether there is too much rain or not enough rain, the relentless propaganda to which we've been subjected issues warnings that these weather events are all the result of climate change that we have caused.

What's frightening about this propaganda is that so many people believe it without question. Those advancing this barrage of lies to which the world has been subjected for nearly forty years know that, as Adolf Hitler stated, "If you tell a big enough lie and tell it frequently enough, it will be believed." And the purveyors of propaganda are banking on the unfortunate tendency of people to forget past predictions that didn't materialize, as described in Chapter 2.

The corrupt media seem to be getting away with pumping out these false stories nonstop. Now, after hundreds of cleverly orchestrated false alerts and tipping points, it seems that some people have more or less accepted the concept that climate change is "here" and is "even worse than predicted".

For example, we read here: "As climate change becomes more severe than originally projected, estimated adaptation finance needs for developing countries are doubling or tripling every few years."[13] An unintended blessing of the COVID-19 scam we've just lived through, with its tyrannical lock downs, mask mandates and vaccine mandates, is that now *many people are finally awake to the fact that they've been lied to incessantly by governments and the media.*

Shortly before his death author Michael Crichton, famous for *Jurassic Park* and the TV series ER, gave a talk in which he compared climate alarmism to a religion for "urban atheists".

> "There's an initial Eden, a paradise, a state of grace and unity with nature, there's a fall from grace into a state of pollution as a result of eating from the tree of knowledge, and as a result of our actions there is a judgment day coming for us all. We are all energy sinners,

> doomed to die, unless we seek salvation, which is now called sustainability."

Unfortunately, the "renewables" being pushed by climate alarmists are anything but sustainable. These "renewables" are destroying the environment (See Chapter 8).

Here is a typical incorrect pie graph of "greenhouse" gases — just one of the 99.9% found by search engines:[14]

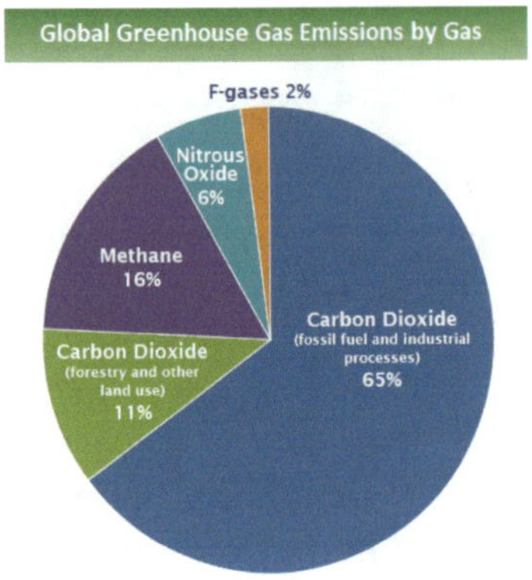

Incorrect! The main greenhouse gas, water vapor, is excluded.

Some of the people promoting this lie may not even realize that the ideas they are pushing are wrong, because the lies being presented are everywhere. You can't escape from them. When preparing this book, I looked for a graph depicting the percentage of various "greenhouse gases". It turned out that 99.9% of all such graphs found online were like this one. They had forgotten water vapor! After putting in the search term, "CO_2, 0.04% of atmosphere," I was finally able to find a pie graph that was correct. Below is a graph of greenhouse gases as they actually are. It was very difficult to find this graph. This is a supreme example of how citizens have been lied to — how completely the students in our schools are being propagandized.

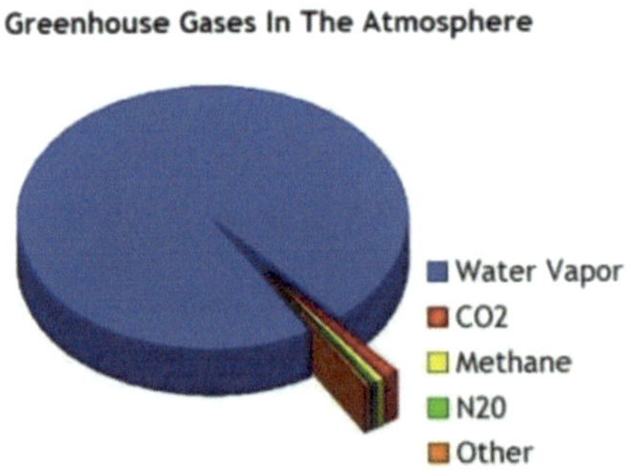

Water vapor is the main greenhouse gas.

The vice president of the US, Kamala Harris, expressed her concern that some young people struggle with a "climate mental health" problem. "They worry about whether to start a family [in view of what's ahead]. They might have to get a job! The stress! How can they adapt the education they have now to their activism?"[15]This is an example of the effect such propaganda, amounting to terroristic threats, has had on our young people.

---

1. Fourth National Assessment,"Volume II: Impacts, Risks, and Adaptation in the United States," https://nca2018.globalchange.gov

2. "An Inconvenient Truth for Al Gore", Michael Dobbs, The Washington Post, 10/12/2007

3. https:junkscience.com/2011/11/suzuki/where/will/santa/live

4. Dr. Patrick Moore, in a March 15, 2019, tweet.

5. Maggie Astor, "The 2017 Hurricane Season Really Is More Intense Than Normal." The New York Times, Sept. 19, 2017

6. Michael B. Sauter, "Cities Around the World that will Soon be Under Water," 247wallst.com, 4/7/22

7. Ibid.

8. Mark Hertsgaard, "Climate Denialism is Literally Killing Us," The Nation, Sept. 25-Oct. 2 issue

9. Philip Bump, "This Map of Sea Level Rise Is Probably Wrong. It's Too Optimistic," https://www.theatlantic.com/politics/archive/2013/11/map-sea-level-rise-probably-wrong-its-too-optimistic/354766/, Nov. 4, 2013

10. Robert Craychik, "Greenpeace Founder: Global Warming Hoax Pushed by Corrupt Scientists 'Hooked on Government Grants'", Breitbart News, March 7, 2019

11. Eric Worrall, "Damage Control: CNN Explains "Global Warming" Does Not Always Mean Warming,"https://www.sgtreport.com/2019/03/damage-control-cnn-explains-global-warming-does-not-always-mean-warming/, March 3, 2019

12. https://www.msn.com/en-us/weather/topstories/10-climate-tipping-points-the-world-needs-to-avoid-at-all-costs/ss-AAY5bWM?bk=1&ocid=msedgntp&cvid=14b514f58c8b40adb444b951031a1919

13. https://www.brainyquote.com/quotes/adolf_hitler_385028

14. Greenhouse Gases, A Student's Guide to Global Climate Change, US EPA

15. Sean Hannity, video clip; Fox News, March 9, 2023

Chapter Eight

Unsustainable "Solutions"

> "The right response to the nonproblem of global warming is to have the courage to do nothing." – Christopher Monckton

The quick-fix "solutions" to climate change offered by the elites running the world are wind power, solar power, and biofuels, multibillion-dollar industries now in full swing. They and their friends are cashing in on this scam to the tune of trillions of dollars. On the day he took office Joe Biden signed an executive order canceling the Keystone Pipeline and multiple oil drilling leases. This started a catastrophic rise in the price of gasoline, heating oil and diesel fuel. Rising energy prices, especially for transportation, cause a significant jump in the prices of other commodities, which is what we're seeing right now.

Wind power is neither green nor sustainable.

You've been told repeatedly by the media — in magazines, TV, the Internet, etc.— that electricity produced by the wind industry is clean, essentially free energy, better for the environment than traditional sources such as coal and natural gas. None of these claims are true.

These wind installations injure, maim, and kill hundreds of thousands of birds and bats each year in clear violation of federal law. The Golden Gate Audubon Society in California reported that the wind farm at Altamont was killing 10,000 birds, including 1,123 birds of prey, each year. Due to a lawsuit, the Altamont wind turbines have had to cease operation for the four months annually that these birds are migrating.[1] Offshore wind turbines have similar impacts on marine birds, and, according to the Union of Concerned Scientists, wind farms located offshore also impact fish and other marine wildlife. [2] Strangely, the numbers and types of birds killed by wind turbines are ignored by wind farm proponents, even though these are substantial.[3] How can this be considered environmentalism?

Currently the construction of an offshore wind farm about 15 miles off the coast of Massachusetts is underway. The foundation pieces for the huge wind turbines, called monopiles, are being driven into the sea floor by pile drivers. Pile-driving noise can deafen, injure or even kill marine mammals. At least fourteen dying humpback whales have recently been washed up on beaches in this area. The people building these projects are fully aware of the damage to marine life that they are causing and will cause. Although killing such creatures is illegal, these companies have been given special permission by the government to injure or kill up to hundreds of whales and thousands of other marine mammals, especially dolphins. "It is hard to see how this hell can be allowed under present laws, such as the Marine Mammal Protection Act and the Endangered Species Act," noted David Wojick of the Heartland Institute.[4]

In May of 2022, Sean Hayes, a NOAA scientist, issued a warning that "the development of offshore wind poses risks to these species [whales]" and that "these risks occur at varying stages, including construction and development, and include increased noise, vessel traffic, habitat modifications, water withdrawals associated with certain substations."[5]

Steve Gray, a former congressional candidate, wondered, " Where are the environmentalists? No outrage. No protests against a Democrat led Government backing offshore windmills. The hypocrisy is deafening. 14 dead whales and counting."[6]

As if this isn't bad enough, hundreds of square miles of forests have been clear cut to provide space for wind farms, causing extensive environmental damage. It's been estimated that wind and solar farms require 400 – 800 times the amount of land to produce the same amount of power as conventional power plants.

Modern wind turbines depend on rare earth minerals mined primarily in China.[7] China has a poor record of environmental stewardship. So the process of extracting these minerals imposes wretched environmental and public health conditions on China's people.

Have you ever seen a wind turbine up close? They are huge. It takes a tremendous amount of material to produce one wind turbine. A typical wind turbine, over 500 ft. (50 stories) tall, contains more than 8,000 different parts, made from steel, fiberglass, cast iron, and concrete. One important component is a magnetic material made from neodymium and dysprosium, rare earth minerals mined almost exclusively in China.

In the Chinese village of Dalahai, where the dumping of waste materials from these occurred, villagers' teeth had begun falling out, their hair turned white at unusually young ages and they suffered from severe skin and respiratory diseases. Children were born with soft bones, and cancer rates had skyrocketed. There were also unusually high rates of cancer, osteoporosis and skin and respiratory diseases reported. The nearby lake's radiation levels are ten times higher than in the surrounding countryside.[8]

As the wind industry has grown, so has the demand for neodymium, which could increase by as much as 700 percent over the next 25 years, and for dysprosium, which could increase by 2,600 percent. An article in *The Daily Mail* stated that every wind turbine contributes to "a vast man-made lake of poison in northern China."[9]

A huge amount of carbon dioxide is generated in manufacturing, transporting and installing wind turbines. The foundation of just one industrial wind turbine contains over 500 cubic meters of concrete, emitting 624,000 kg (1,372,800 lbs) of carbon dioxide to produce. This doesn't include the CO_2 emitted from transporting this concrete to the job site. This fact reveals the hypocrisy of those who claim that wind turbines are good for the environment. Considering the number of animals these monstrosities kill and the pollution created during their manufacture, how could anyone who truly cares about the environment approve of their use?

The decommissioning [scrapping] of wind turbines that have worn out after their 10 – 15 years of useful life are not included in any CO_2 cost-saving calculations, making emission claims a farce. The huge fiberglass blades cannot be recycled. They are just buried. Many older wind farms that have outlived their usefulness have simply been abandoned, because properly disposing of them would be prohibitively expensive.

To be competitive with other sources of power, wind farms require huge government subsidies paid for by taxpayers. They also need vast stretches of land. Rockefeller University professor Jesse Ausubel has calculated that meeting New York City's electricity needs alone would require blanketing the entire state of Connecticut with these towering turbines.[10]

Because the wind doesn't blow all the time, every wind turbine must be backed up by expensive coal or gas power plants that sit idle when the wind is blowing at just the right speeds. These conventional sources must then kick in whenever the wind dies down in order to prevent factories, schools, offices and homes from experiencing blackouts or brownouts.

Richard S. Courtney explained,

> "Windfarms for power generation provide intermittent power, so they merely displace thermal [fossil fuel] power stations onto standby mode or to operate at reduced efficiency while the thermal power stations wait for the wind to change. They make no significant reduction to pollution, because thermal power stations must continue to use their fuel and produce their emissions while operating in standby mode with reduced efficiency. Their emissions from operating at a low output are actually increased."[11]

In Germany utilities are forced to buy "renewable" energy at several times the cost of conventional power. In the United Kingdom, The Telegraph reported that, rather than providing cheaper energy, wind power costs the electric companies

£50 per megawatt-hour, compared to £15 for conventional power - over three times the cost.

The same is true of any electrical grid. Wind turbines respond only to the wind, so their contribution to the supply of electricity on the grid is essentially inconsistent and unreliable. The wind may not be blowing when demand is high. When the demand is low, it could be blowing just right. Wind turbines don't operate when the wind speed is below six meters per second and must be shut down if it is greater than 25 meters per second (to protect blades and bearings).

This is the dirty little secret behind why Big Oil loves wind power. Fossil fuel generating plants will always be needed to back them up, and continually powering up and down is much more wasteful of energy than constant operation.

Broken wind turbine

In February, 2019, John Hinderaker stated, "Here in Minnesota, we are enduring a brutal stretch of weather. The temperature hasn't gotten above zero in the last three days, with lows approaching -30°. And that is in the Twin Cities, in the southern part of the state. Yesterday central Minnesota experienced a natural gas "brownout," as Xcel Energy advised customers to turn thermostats down to 60 degrees and avoid using hot water. Xcel put up some customers in hotels. Why? Because the wind wasn't blowing."[12]

Professor Randy Simmons added,

> "Over the past 35 years, wind energy – which supplied just 4.4% of U.S. electricity in 2014 – has received $30 billion in federal subsidies and grants. These subsidies prevent people from learning the uncomfortable truth of just how much wind power actually costs. They transfer money paid by taxpayers [without their knowledge] to wealthy wind farm corporations, many of which are units of foreign companies."[13]

Several studies have placed doubt on the value of wind turbines. A 2011 study at Purdue University and a 2016 study in Scotland found that wind farms were actually increasing surface temperatures. A 2018 study concluded that wind power in the US could warm surface temperatures by 0.24° C., nearly one-fourth of the global temperature increase experienced since 1800.[14]

The Institute for Energy Research reported that the actual cost of wind energy is about twice that estimated by the Department of Energy. Because the useful life of a wind turbine is about 10-15 years, not the optimistic 30-year figure provided by the government, claims by the wind industry that wind power is becoming cost competitive with fossil fuel-generated power are absolutely false.[15] If wind power was less expensive than — or even merely cost-competitive with — coal and natural gas energy, these subsidies and mandates would be unnecessary.[16]

Added to these problems are the adverse health effects of living anywhere near these wind turbines. The incessant grinding noises they make have been known to cause tinnitus [ringing in the ears], vertigo, panic attacks, migraine headaches, sleep deprivation, and even heart disease, according to groundbreaking research published by Dr. Nina Pierpont, a leading New York pediatrician.

Wind turbines kill millions of birds and bats.

Dr. Pierpont studied the symptoms of people living near wind turbines in the United States, the United Kingdom, Italy, Ireland and Canada. Her findings have led her to confirm what she has identified as a new health risk: wind turbine syndrome (WTS).[17] This is the disruption or abnormal stimulation of the inner ear's vestibular system caused by turbine infrasound [low-frequency noise], the most distinctive feature of which is a group of symptoms which she calls visceral vibratory vestibular disturbance, or VVVD. They cause problems ranging from internal pulsation, quivering, nervousness, fear, a compulsion to flee, chest tightness and tachycardia [increased heart rate]. Turbine noise can also trigger nightmares and other disorders in children as well as harm cognitive development in the young.[18] If wind turbines are affecting people this way, what are they doing to farm animals and wildlife?

Locating wind farms offshore or in unpopulated areas is not a practical solution. This is because electricity must then be transported long distances to reach the cities, where the demand is greatest. The farther electricity has to travel from the producer to the user the more is dissipated en route, wasting power.

Offshore wind turbines

Winds are stronger and more constant in offshore sites than on land installations. However, expensive mountings must then be built to prevent large ocean waves from dislodging them from their moorings.

In February of 2021 Texas and nearby areas were hit with the worst winter storm in over a century, with several consecutive days of freezing temperatures. Hundreds of wind turbines iced up and failed to operate, causing a nightmare for ERCOT, the electricity provider. Millions of people were without electricity for several days. An estimated 200 people died. The cost of these unreliable wind turbines was over thirty billion dollars: over $1,000 for every man, woman and child in Texas.

A year later, on February 24, 2022, Russia invaded Ukraine. Despite all the hype from the Paris Accord about converting from fossil fuels to renewables, Europe is still heavily dependent on Russia for its energy needs. Europe produces 3.6 million barrels of oil per day but uses 15 million barrels per day. Russia supplies 20% of Europe's oil, 40% of its natural gas and 20% of its coal.

Since the Biden regime canceled the Keystone Pipeline and numerous oil leases, the United States is no longer energy independent. The Keystone Pipeline alone would have delivered 830,000 barrels of oil a day to the US.

In view of the above, the massive subsidies of around 40% of the cost of each wind turbine provided to wind power companies in the US have proved to be a tremendous waste of tax dollars. Industrial wind farms are ridiculously expensive and destructive to communities and local environments. Fragile ecosystems are destroyed. Birds and bats are killed. Animal habitats are disrupted, and wildlife is driven from the area. Noise and vibrations cause human health problems, including sleep problems, headaches and nervous disorders. Home values and real estate prices are reduced, with some properties becoming impossible to sell. Energy from wind power is not a viable alternative to carbon-based energy production.

The Truth About Solar Energy

Photovoltaic cells, made of semiconducting materials such as silicon, convert sunlight directly into electricity. We use them for all sorts of jobs such as powering watches and calculators and road signs that light up. When photons of sunlight strike these semiconducting materials, they knock electrons loose from their atoms. These electrons are then directed through the semiconducting material to a light or motor.

Under ideal conditions, when there is a cloudless sky at noon, at an angle of 90 degrees to the direction of sunlight streaming in from the sun, one square meter of the panel will produce about 1000 watts, or one kilowatt, of power. This is known as Watt Peak.

But often it is rainy or cloudy. Only when the sun is highest in the sky at noon on a cloudless day will you have ideal conditions. The solar panel must face south in the northern hemisphere (and north in the southern hemisphere). It must be at the same angle as the latitude of the installation. During winter months, when the sun doesn't get as high in the sky, and during morning and evening hours, the wattage — and the efficiency — of the photovoltaic panel drops off. And then, of course, it is nonexistent at night when the sun isn't shining. This is why a typical solar panel has an average efficiency of only about one-sixth (17%) of its peak rated power. And the efficiency becomes even lower over time as the solar panel starts to wear out.

The average cost of a residential solar panel installation in 2022 ranged between $18,000 and $20,000. Applying the tax credit paid for by taxpayers, you'd save about 26% of this cost. So you'd actually pay between $13,320 and $14,800. This works out to between $2.75 and $3.35 per watt, plus electrical and roof work, trenching, batteries, etc. Through our taxes, we are helping others finance the construction of their solar projects.

A YouTube video presenter claimed that his system, rated at 4800 watts, was providing 4800 watts of electrical power. He used this figure to compute the time needed to recover his original expenditure as only 7.1 years. This is a mistake many people make, because they don't understand that the efficiency of such a system will never exceed about 17% of the rated power (peak wattage). He will be lucky to get 816 watts (17% of 4800 watts). The time needed to recover the cost of this system would thus be about 42 years, not 7.1 years. The average life expectancy of such a system before it has to be scrapped is about 21 years. So even with the 26% tax subsidy a person will never recover the investment he made in it. And this doesn't include the substantial cost of having it scrapped after it wears out. So solar power is not "free", as advertised. In fact, consumers investing in this technology gets back a return of about 12 cents for every dollar spent.

Large solar projects consume a tremendous amount of land area. To produce enough electricity for a practical large-scale system a project needs several acres of land for each megawatt of power. Solar plants need about 400 - 800 times as much land as natural gas or nuclear electrical generators for the same amount of power. These huge solar farms impinge on and threaten natural ecosystems. How could any real environmentalist approve of such systems?

One type of commercial solar electricity generating plant is a concentrated solar power plant, also known as solar thermal. In this type of arrangement each of an array of thousands of mirrors surrounding a receiving tower is positioned at the precise angles needed to reflect sunlight onto the tower. The Gemasolar power plant in Spain consists of 2,650 such mirrors and occupies 457 acres of land. This produces enough power to supply 25,000 homes. Sierra Suntower, a plant in the Mojave Desert in California, has two towers and 24,000 mirrors, is rated for 5 MW (million watts) providing energy for 4,000 homes.

You can learn more about these solar projects by accessing solarpaces.nrel.gov, which provides a list of projects, their cost to construct, capacity, location and current operational status. Because the sun's position in the sky is constantly changing, due to Earth's rotation and changing seasons, each of the mirrors depicted here must have its own motor to change its position in order to keep the reflected sunlight focused on the tower. The central tower contains water or molten salt at a temperature of about 500°C. circulating through pipes. This molten salt heats water to steam, powering an electrical generator.

A thermal solar farm.

The advantage of this type of arrangement is that some of the hot molten salt can be kept in an insulated tank to be used for heating water at night or on cloudy days. This type of operation, called a thermal battery ["thermal" meaning "heat"], doesn't need a battery backup for times when the sun isn't shining as do traditional photovoltaic plants.

These plants are extremely expensive. They take up a tremendous amount of acreage and destroy wildlife habitats. Maintenance costs to keep such a plant running are high, because each mirror must be at peak reflectivity and exactly the right angle. Solar plants are often located in deserts where cloudy days are rare. But two problems arise from this. First, water, scarce in a desert, is required to operate such a system. Deserts are usually far from population centers. A great deal of electric power is lost in transporting it over long distances.

Feeding solar energy into the electrical grid is tricky. Peak solar generation occurs in the early afternoon when the sun is highest in the sky. Peak demand for electricity doesn't occur until the evening hours. This mismatch between peak output and peak demand has been dubbed the "duck curve". In the evening, as solar generation decreases due to the sun's position in the sky, system operators must use coal, natural gas or nuclear backup systems to prevent power outages and brownouts. Experience has shown that if the input to the grid from a solar system exceeds about 6%, there is actually a negative effect on the power grid that overloads it.[19]In this case production must be curtailed. This electrical energy cannot be stored in batteries, because battery technology hasn't developed to the point where batteries can hold more than a few hours of electrical power.

In 2016 California ISO (Independent System Operator) was forced to curtail over 3.08 GWh (billion watt hours) of electricity, which went to waste because supply exceeded demand. As ISO put it,

> "While curtailments are currently representing only a fraction of overall load, the issue is expected to intensify in the coming years as the state strives to achieve its 50-percent renewable target...Curtailing renewables is counterintuitive to California's environmental

> & economic goals. It reduces the output from the renewable plants in which the state has invested, and could result in overbuilding renewable plants to ensure that the state meets its 50% renewable mandate. Overbuilding the state's electrical system isn't financially sound." [20]

Large solar projects receive government subsidies of at least 30% of their cost, paid for by the taxpayer. Many taxpayers have no idea that they are actually footing a substantial part of the bill for these projects.

Many states are now imposing penalties on power companies that fail to include a specified percentage of the power they produce from these expensive wind and solar sources.[21] There is a double whammy here. First an additional burden is put on taxpayers to pay the subsidies. Second, because wind and solar power systems are so expensive, the power companies have no other option but to pass this increased cost of electricity onto their customers. The cost of electricity has more than doubled in Great Britain, Germany and many other places.

The construction of photovoltaic panels also causes great harm to the environment and human rights. An estimated 70% of the World's photovoltaics are produced in China, much of it with slave labor (in Uighur Muslim concentration camps). The New York Times reported in 2014, "Although China may be a cheaper place than Europe for producing solar panels, the savings come at a higher cost to the environment." In their analysis scientists added together all the costs involved in production: the mining of raw materials, transportation, and the power requirements for production of these panels.

"Sometimes the environmental costs of solar panel production can be lost among the drive to encourage the development of clean energy," declared Huang Xianjin, a Nanjing University professor who studies land use. Huang and two other professors submitted a letter to the journal, *Nature*, arguing that China needed to take significant action to offset the environmental damage the solar industry is causing there.[22]

A solar panel manufacturer in Zhejiang Province halted its operation after local residents complained about serious air and water pollution caused by their production. The Luoyang Zhonggui High-Technology Company produces a highly toxic byproduct of polysilicon used in solar panels called silicon tetrachloride. "The land where you dump or bury it will be infertile. No grass or trees will grow in the place...It is like dynamite - it is poisonous, it is polluting. Human beings can never touch it," said Ren Bingyan, professor at Hebei Industrial University in China. The irony is that solar is touted as green energy.[23] Because of the increasing demand for solar energy, the price of polysilicon soared from $20 to $300 per kilogram in five years. Dozens of new polysilicon plants are riding the crest of China's prosperity. The combined capacity of these new factories is estimated at 80,000 tons, over twice the amount of polysilicon produced in the rest of the world.

Due to lax standards, these companies are cutting corners with hazardous waste disposal. Although polysilicon producers in other parts of the world recycle this compound, the Chinese companies prefer to dump it to avoid the high cost of recycling. They've failed to install the technology needed to prevent pollutants from getting into the environment.

Then there is the problem of the disposal of worn out solar panels. Michael Shellenberger expressed his concern about environmental hazards resulting from discarded solar panels. Typically they contain cadmium, lead and other toxic chemicals which can't be extracted without taking the whole panel apart.[24] Such panels left in landfills break apart and release toxic waste into the ground or bodies of water.

In a 2016 study Electric Power Research Institute concluded, "Solar panel disposal in "regular landfills [is] not recommended in case modules break and toxic materials leach into the soil." Over a period of time rainwater will wash cadmium out of solar panels and into the environment. This fact was brought up when local people In Virginia rejected a proposal to construct a 6,350 acre solar farm in Spotsylvania County. Sean Fogarty of Concerned Citizens of Fawn Lake stated, "We estimate there are 100,000 pounds of cadmium contained in the 1.8 million panels."[25]

There have been some promising advances in solar electricity generation. But many more years of research and development are needed to make solar energy cost competitive and environmentally safe. In order to eliminate the need for backup by coal and natural gas, batteries, which are now totally inadequate, must be improved.

At this time the battery technology needed to store this energy is woefully inadequate. "The reason winter is cold is because of a lack of solar energy. The sun is low in the sky, days are short, and it is cloudy much of the time. Yet climate alarmists want people to be dependent on solar energy being generated under those conditions for their survival. They imagine that there is some storage technology which can store huge amounts of energy for long periods of time when the sun isn't shining or the wind isn't blowing."[26] The problem is that these haven't been found yet.

The Green New Deal and other carbon tax schemes anticipate having all of these problems solved within a decade or so. This is an impossible pipe dream entertained only by uninformed politicians and others who don't have a clue about the science or the technology. It appears that the Biden administration is forcing everyone to do the impossible.

Seven American refineries, which formerly processed 806,000 barrels of petroleum per day, have gone offline. Some simply shut down while others are being converted to biofuels like ethanol to meet the demands of the environmental lobby. "These losses have left the US with 124 operating refineries. That's 10% fewer refineries for gasoline, diesel and jet fuel than in 2016, even as demand has continued to increase."[27] When people complain about the high cost of gasoline and diesel, they are told, "Go buy an electric car."

Energy blackouts are predicted for the near future because of the installation of "renewables" and the cancellation of dependable power sources. Wind and solar energy farms require either massive, utility scale battery farms or backup from reliable hydrocarbon powered generators, which costs a fortune. This has been compared to a person jumping out of a plane and then wondering if he is wearing his parachute.

Grossly Over-Rated Biofuels

There's nothing new about biofuels, or fuels derived from plants such as trees and corn. They've been used since the dawn of history, when people started burning them for cooking and heating their homes. Later, with the mining of coal and discovery of petroleum in the late 1800s, biofuels lost their popularity, especially in developed nations.

Yet, the US Energy Information Administration reported that in 2015 a full 43% of "renewables" were biofuels: trees, crops, manure, and trash burned to produce electricity, heat and transportation.[28]

Ethanol Storage Tanks

One type of biofuel, ethanol (ethyl alcohol), can be made from either corn or cane sugar. This is the same kind of alcohol found in alcoholic drinks. During the two world wars in the first half of the 20th century, biofuels were made from several different substances, including potatoes. But when huge supplies of crude oil became available in Texas and Oklahoma, gasoline became a less expensive alternative to biofuels.

In the 1970s the OPEC (Organization of Petroleum Exporting Countries) nations arbitrarily cut their oil exports to raise the price of oil, creating fuel shortages. This caused governments to shift their interest back to biofuels. Part of the reason for this, they said, was that petroleum was a scarce, nonrenewable resource. With more oil fields being discovered every year, we now know that this isn't the case at all. There is now an estimated 14 trillion barrels of petroleum reserves waiting to be extracted from the ground. In comparison, since petroleum was discovered in 1859, only about one trillion barrels, or 8% of known reserves, have been used.

A group of Democrats led by Congressman Henry Waxman made a deal with the corn lobby. In exchange for its financial support of their political campaigns, they engineered a mandate to blend corn derived ethanol with gasoline as an amendment to the 1990 Clean Air Act.[29] Advertised as another great source of alternative, clean energy to solve the world's energy problems, ethanol started being added to gasoline in amounts up to 15% without the consent of the consumer. It is now difficult in many parts of the United States to find gasoline stations that offer the option of pure gasoline without ethanol.

But gasoline containing ethanol quickly destroys boat and lawn mower engines. Many consumers have had the unfortunate experience of having their lawn mowers and rototillers destroyed because they were never warned about the damage the gasoline/ ethanol mixture causes in these small engines. It damages rubber and plastic components of fuel systems, especially in older models built of materials that aren't resistant to this type of damage. This includes gaskets, which become brittle, two-cycle carburetors and primer and return lines, which start leaking after a while.

If ethanol isn't burned within sixty days, it will literally decay.[30] This is because ethanol fuels are hygroscopic. They attract water. Because of this, fuel separation can occur and damage the engine's carburetor. Fuel which has decayed in this way contains varnish and sludge deposits. Fuel that is 15% ethanol, called E85, cannot be used in a vehicle unless its engine has been specially modified.

Ethanol is now a worldwide phenomenon. With biofuel mandates now in over 60 nations, the competition between ethanol and food for agricultural land has become a moral issue. Groups like Oxfam and the Environmental Working Group oppose biofuels, because they create food shortages, raising the price of food.

Ethanol production is a huge business. Millions of acres of virgin prairie and wetlands have been plowed up for the first time ever. The increased use of nitrate fertilizers in growing these crops has caused massive runoffs of nitrates into creeks and rivers, contaminating water supplies in Midwestern states. Cleaning up this water has become very costly. Over a third of all the corn produced in the United States has been going into the production of ethanol.[31] 4.2 million acres of additional land, an area double the size of Yellowstone Park, is now being used for biofuels. Native prairie never before used for agriculture has been tilled up. In 2021 fifteen billion gallons of ethanol were produced in the US.[32] Yet at the same time almost eight hundred million people in the world are malnourished. And with the war in Ukraine, food shortages are expected. The situation is expected to get a lot worse.

"The UN's Food and Agriculture Organization (FAO) reported a 45% increase in the world food price index during the 2008 food crisis. Wheat prices increased by 130% relative to 2007 levels. Similarly, soy prices went up by 87%, rice prices by 74%, and maize prices by 31%." Causes of this crisis include the growing demand for biofuels.[33] Now lobbyists for this industry are pushing for legislation that would raise the maximum percentage of ethanol allowed in gasoline from 15% to 30%.[34]

The University of Minnesota and the University of Wisconsin jointly studied changes in land use across the US resulting from the ethanol industry. The leader of this study, Chris Wright, expressed his concern that native grassland was being converted to agricultural land, which has a negative impact on the environment. The Environmental Protection Agency noted in 2011 that this practice causes increased soil erosion and loss of wildlife habitat. These are unfortunate and unintended consequences of ethanol production.

A gallon of gasoline delivers 116,900 BTU's [British Thermal Units] of energy, whereas a gallon of ethanol delivers only 76,000 BTU's of energy, about 2/3 the energy of gasoline. So a vehicle gets fewer miles per gallon of gasoline/ethanol blend as it does on pure gasoline. Yet it costs $3.95 to produce one gallon of ethanol from corn. And now, because of the current, planned shortage of oil in the United States, Biden has just announced that the production of ethanol will be increased.

Adding up the energy costs of corn production and its conversion to ethanol, 131,000 Btu are needed to make one gallon of ethanol. "Put another way," Professor David Pimentel of Cornell University stated, "about 70 percent *more* energy is required to produce ethanol than the energy that ethanol contains. Every time you make 1 gallon of ethanol, there is a net energy loss of 54,000 Btu."[35] He also reported, "The use of corn for ethanol has led to major increases in the price of U.S. beef, chicken, pork, eggs, bread, cereals and milk – a boon to agribusiness and a bane to consumers."

The Director-General of the UN Food and Agriculture Organization, Jacques Diouf, reports that using 22 pounds of corn to produce one gallon of ethanol is already causing food shortages.[36]The price of food has also increased dramatically due to unusually cold weather ruining crops.[37] Government subsidies for the corn industry include tax breaks, grants, loans, and loan guarantees. This subsidy averages $1.65 per gallon, coming right out of your taxes.

"It is now clear that the federal corn ethanol mandate has driven up food prices, strained agricultural markets, increased competition for arable land and promoted conversion of uncultivated land to grow crops. In addition, previous estimates have dramatically underestimated corn ethanol's greenhouse gas emissions by failing to account for changes in land use."[38]

It is immoral to use agricultural land that could be growing food for fuel. The revolution in Egypt called "Arab Spring" started with riots caused by food shortages that were directly caused by this newly instituted practice.[39] But then, believe it or not, "human-caused climate change" was blamed for these rising food prices. Jean Ziegler commented, "It is a crime against humanity to convert agricultural productive soil into soil which produces foodstuff that will be burned as biofuel."[40]

"We thought and hoped it would be a climate solution and reduce and replace our reliance on gasoline," commented Tyler Lark, a researcher with the Nelson Institute for Environmental Studies at the University of Wisconsin-Madison, and lead author of the study. "It turns out to be no better for the climate than the gasoline it aims to replace and comes with all kinds of other impacts." Dr. Lark stated that ethanol puts 24% more carbon dioxide into the environment for the same amount of power than gasoline does. [41]

Biodiesel fuel is a blend of about 80% petroleum diesel and 20% biodiesel, which is produced by treating soybean or another oil with methanol and a catalyst. This industry also receives substantial government subsidies through the Biomass Crop Assistance program (BCAP). Producers are reimbursed up to 50% of their cost with payments extending up to five years for herbaceous feedstock and 15 years for woody plants.

Section 9003 of the Biorefinery Assistance Program guarantees loans to producers for the construction, development, and retrofitting of commercial-scale biorefineries.[42] These fuel sources can be competitive only through these subsidies, which bilk the taxpayer. Since there is a surplus of petroleum and natural gas, there is no need for ethanol and other biofuels. The government policies responsible for this scam represent the epitome of corruption.

In the 1970s, when OPEC, the Organization for Petroleum Exporting Countries, generated a false crisis by cutting their exports, we were told that the world would be running out of oil in twenty years. This was also falsely predicted by the Club of Rome.

But now Michael J. Economides, one of the world's foremost experts on the geopolitics of energy, has stated, "approximately 6-8 trillion barrels of oil from conventional resources, with the same or more from unconventional [tar sands, shale oil etc.] resources are believed to be 'in-place' and as yet untapped," with a total of 12-16 trillion barrels in place.[43] But since the world started using oil, only about one trillion barrels of petroleum have been used. So the big question is: with all these available petroleum resources, why is there a need to use valuable agricultural land and rob the earth of virgin prairie and tropical rain forests to produce ethanol?

Climate alarmists have been concerned about a drought in the western United States, which they say is due to climate change. Yet 20%-30% of the water in the Colorado River, west of the Continental Divide, is extracted for growing corn, much of which is used to produce ethanol. So the climate change scam is contributing to the loss of water in this region.[44]In California fresh water is being dumped into the Pacific ocean. "Since the beginning of January [2023] a series of atmospheric rivers has disgorged trillions of gallons of much-needed moisture across drought-stricken California, but only a small fraction of that water has so far made it into storage. In the delta — the heart of the state's vast water system — nearly 95% of incoming water - trillions of gallons – has been allowed to flow into the Pacific Ocean, according to data from the U.S. Bureau of Reclamation."[45]

Using Wood to Produce Electricity

In some states forests are now being clear cut, and the trees ground up and burned to produce energy. They are replanted with just one or two species of trees, reducing biodiversity from 10-20 species to just one or two species of trees. Devastating to the environment, this practice disrupts nature's balance. Forests owned by the state of Massachusetts "are being logged at drastic, nearly clear-cut levels and at unsustainable rates."[46]

A clearcut forest waiting to be shipped

"Biomass is an appealing 'easy out' for wasteful energy policies and land barons hungry for quick cash. This sweetheart deal for big landholders is being sold as 'carbon neutral', even though burned wood releases 1.5 times as much CO_2 as coal per unit of mass. Taxpayers are lured in by promises of cheap energy and reduced taxes on the myth that this practice is sustainable or healthy for forests, people or the climate."[47]

To comply with new regulations calling for the use of "renewables" in the United Kingdom, Drax, its largest electrical generation plant, is now burning wood shipped 3000 miles from forests in North Carolina. Under European Union rules, wood imported from outside the EU is considered to be a renewable, zero-carbon fuel source. Going along with that, the United States Congress declared nearly all forest biomass in the United States to be carbon free. But declaring a substance "carbon free" doesn't make it carbon free.

Burning wood releases several toxic organic compounds, including some that produce acid rain. Harmful emissions include carbon monoxide (a deadly gas not to be confused with life-giving carbon dioxide) formaldehyde, benzene and dioxins, all of which are carcinogens. Some particulates released have been known to cause asthma and other respiratory diseases as well as cancer.

When these biofuels are burned, just as with gasoline and coal, carbon dioxide is released.[48] Biofuel advocates claim that the combustion of these plants is "carbon neutral" because the trees absorbed carbon dioxide while they were growing. But it takes 40 to 70 years to replace a forest. Since they say there are less than ten years left to save the planet, the newly planted trees won't even have time to reach maturity.

Even worse, when these trees are cut down, the soil microbes located underneath the cut trees release just as much carbon dioxide into the atmosphere as does the combustion of the trees themselves. So when a tree is cut down and burned, nearly *three times* as much carbon dioxide is released into the atmosphere as an equivalent amount of coal.[49]

In the "green" bureaucrats' flawed energy accounting system, the practice of burning wood is considered superior to the combustion of coal that could have been mined right in the UK. The energy required to cut down this forest, convert the wood into pellets and then transport the huge quantity of wood needed to run this immense facility has obviously never been considered. Yet this practice has been hailed as a great achievement by people who think they are "green".[50]

In setting the goal of doubling Europe's 2015 renewable energy levels by 2030, the European Union decreed that cutting down forests to be burned in electric power plants instead of using coal or natural gas qualifies as zero-carbon, renewable energy. They did this after nearly 800 scientists warned them that this policy would actually increase CO_2 emissions.[51] The Renewable Energy Directive (RED) is now finalized. It is obscene that a huge quantity of wood – as much as all the other uses of wood worldwide – will now, due to such brainless policy decisions, be required to meet just a tiny fraction of Europe's energy requirements. How could this practice ever be construed as helping the environment?

Wood burned in an electric power plant emits up to three times as much CO_2 as natural gas. This is because of wood's carbon bonds, water content, and the lower temperatures at which wood burns. It's been estimated that replacing coal and natural gas with wood could result in a CO_2 load 2-3 times greater than carbon-based fuels by the year 2050.[52] After the forest is cut down, the trees will no longer be there to absorb the CO_2 they took in from the atmosphere when they were alive. Whether or not the extra carbon dioxide will affect the climate, this regressive action shows a significant flaw in the knowledge and judgment of bureaucrats making such foolish decisions.

This harebrained decision by the European Union sent an adverse signal to the rest of the world. At the last global climate conference countries with tropical rain forests - including Indonesia and Brazil - jointly stated their goals "to increase the use of wood...to generate energy as part of efforts to limit climate change."

---

1. Jonathan Benson, "'Green' Debacle, NaturalNews.com, Nov 24, 2011
2. /www.ucsusa.org/clean-energy/renewable-energy/environmental-impacts -wind-power#.WqAYMM9x0xA
3. "Wind Farms Raise Temperatures at the Surface Level", Institute of Energy Research, March 20, 2023
4. David Wojick, "Whale Hell Looms in Massachusetts", Feb. 18, 2023, The Heartland Institute
5. Peter Partoll, Spate of Whale Strandings Near Offshore Wind Projects Becoming Impossible to Ignore – But Feds Still Deny Link, The Western Journal, March 4, 2022
6. May 3, 2023
7. http://instituteforenergyresearch.org/analysis/big-winds-dirty-little-secret-rare-earth-minerals/
8. http://www.dailymail.co.uk/home/moslive/article-1350811/In-China-true-cost-Britains-clean-green-wind-power-experiment-Pollution-disastrous-scale.html
9. https://www.instituteforenergyresearch.org/renewable/wind/big-winds-dirty-little-secret-rare-earth-minerals/

10. http://nucleargreen.blogspot.com/2008/02/renewable-and-nuclear-heresies.html

11. Richard S Courtney, Oct. 26, 2006, Lecture and White Paper presented to the Institute of Mining & Mechanical Engineers

12. John Hinderaker, "Why "Green" Energy is Futile, in One Lesson," Thecoachstream.wordpress.com, Feb. 1, 2019

13. Randy Simmons, "What's the True Cost of Wind Power?" April 11, 2015, Newsweek Magazine

14. Op. cit., Institute of Energy Research

15. http://instituteforenergyresearch.org/press/study-wind-solar-5x-costly-existing-coal-nuclear/

16. cfact.org/2018/05/26/environment-florida-misleads-voters-on-renewable-energy-costs/?mc_cid=2a94bd2c4c&mc_eid=4e33dab1e3

17. http://www.windturbinesyndrome.com/wind-turbine-syndrome/

18. Ibid.

19. The Solar Value Cliff: the Diminishing Value of Solar Power", Aug. 2017, Institute for Energy Research

20. "Fast Facts, California ISO, no date

21. http://dailycaller.com/2016/09/22/the-solar-industry-cashes-in-on-government-subsidies/

22. https://chinadigitaltimes.net/2014/06/chinas-solar-panel-production-comes-dirty-cost/

23. Ariana Eunjung Cha, "Solar Energy Firms Leave Waste Behind in China,"The Washington Post, March 9, 2008

24. Michael Shellenberger, "If Solar panels Are So Clean, Why Do They Produce So Much Toxic Waste?, Forbes Magazine, May 23, 2018

25. https://www.forbes.com/sites/michaelshellenberger/

26. Heller, op. cit.

27. https://hotair.com/jazz-shaw/2022/05/23/its-not-just-electricity-were-heading-for-shortages-of-diesel-jet-fuel-and-gasoline-n471325

28. Josh Schlossberg, "[Study] Biofuels Increase Net Carbon Emissions", Aug. 26, 2016, www.thebiomassmonitor.org

29. William O'Keefe, "The Real Reason We Have Ethanol in Our Gas", Fee.org, Nov. 26, 2016

30. http://b3cfuel.com/resources/education/the-problem-with-ethanol

31. David Pimentel, http://hir.harvard.edu/article/?a=1911

32. https://www.statista.com/statistics/281494/us-fuel-ethanol-production/

33. “High Food Price Crisis,” www.theoaklandinstitute.org

34. Daniel Strohl, “EPA proposes increasing ethanol content in fuel to 30 percent,” https://www. hemmings.com/blog, May 17, 2013

35. https://www.organicconsumers.org/scientific/70-percent-more-energy-required-make-ethanol-actually-ethanol-cornell

36. “Tucker Carlson Tonight”, June 10, 2022

37. “Other Crop Losses Globally Due to Cold and Mini Ice Age Climate Intensification,” May 3, 2017, www.oilseedcrops.org

38. “Ethanol’s Broken Promise: Using Less Corn Ethanol Reduces Greenhouse Gas Emissions”, May 29, 2014, www.ewg.org

39. “Did Food Prices Spur the Arab Spring?” https://www.pbs.org/newshour/world/world-july-dec11-food_09-07, Sept. 7, 2011

40. “UN Right to Food Rappeuteur,” 2007

41. Georgina Gustin, "Corn-Based Ethanol May Be Worse For the Climate Than Gasoline, a New Study Finds", InsideClimateNews.com, 2/16/21

42. AR. G. Cachat, “Massachusetts: the Hoax of Biomass and Modern Forestry”, Aug. 6, 2016, www.the biomassmonitor.orglternative Fuels Data Center, U.S. Dept. of Agriculture

43. Peter C. Glover and Michael J. Economides, Energy Climate Wars, Continuum Intl. Publishing Group, 2010, p. 79

44. Tom Nelson podcast #13, interviewing Tony Heller, Sept. 6, 2022

45. Hayley Smith, "Environmental rules stoke anger as California lets precious stormwater wash out to sea," Los Angeles Times, Jan. 20, 2023

46. Ibid.

47. Ibid.

48. Josh Schlossberg, Ed., thebiomassmonitor.org

49. Josh Schlossberg, Ed., thebiomassmonitor.org

50. Rupert Darwall, “Tesco and the Great Green Scam,” https://blogs.spectator.co.uk/2017/05/tesco-great-green-scam/

51. Searchinger, et al., op. cit.

52. Ibid.

Chapter Nine

Battery-Powered Vehicles

> "All the major parties in Europe fear people are at the risk of freezing to death, like what happened in Texas in March of 2021, due to widespread energy outages. But even if human casualties can be avoided, energy poverty becomes an ever more realistic prospect for large parts of the European populace, a risk driven mainly by ideological decisions that were based on very little scientific evidence." — Ralph Schoellhammer, Asst. Professor, Vienna University

Battery powered cars were popular in the early 1900s. Now they are being promoted as a solution to climate change. But whether or not electric vehicles actually eliminate any carbon dioxide emissions depends on the ***source*** of the electricity being used to charge their batteries. If that electricity was generated by coal or natural gas — as most power plants in the US are — there is no benefit to driving an electric vehicle. A study at the University of Michigan concluded that an electric vehicle recharged by electricity from a coal-powered plant generates the same amount of carbon dioxide as a conventional car getting 29 miles per gallon.[1]

Electric bus fire, Shreveport, Louisiana

Only electricity generated by a "renewable," such as solar or wind power could ever claim to reduce CO_2 emissions.

A study by the Union of Concerned Scientists revealed that building an electric vehicle generates 15% to 68% more carbon dioxide than a conventional vehicle, because of the intensive process of making its battery, the components of which are not renewable.[2]

When people complained that gasoline was selling for $5 per gallon, Senator Stabenow of Michigan responded, "We all need to buy an electric car so we won't have to buy gasoline and be dependent on the whims of the oil companies." But currently they are not practical for long distance driving due to range requirements and the absence of recharging stations. A recent study[3] found that, in the fourth quarter of 2022 the cost of driving some electric vehicles was higher than similar gasoline-powered cars. The cost of fuel for a mid-priced internal combustion engine (ICE) car was $11.29 for every 100 miles traveled. This was $0.31 cheaper than the amount paid by drivers in mid-priced electric vehicles being charged mostly at home. This occurred because gasoline prices were getting cheaper and the cost of electricity was increasing at the same time.

At the present time only about 2% of vehicles in the United States are battery powered. To reach 100% fifty times more battery-powered cars would put an impossible strain on the grid, which is already being threatened by intermittent wind and solar power. Where do the bureaucrats who are running the show think all that extra power will come from? Can the grid really handle 100% of vehicles being EVs? The average annual mileage driven by the 230 million licensed drivers in the US is 13,500 miles. With 33.7 kilowatt hours (kWh) of electrical energy equivalent to 1 gallon of gasoline, this amounts to 4.1 trillion kWh of energy needed, an astronomical figure.

courtesy of businessinsider.com

In the summer of 2022 Tesla drivers in Texas were told not to charge their car batteries unless they had no other choice, due to the strain on the electrical grid (that they claim was due to a heat wave). But summers in Texas have always been hot. 100-degree summertime highs have been common since the beginning of temperature records. Texas officials should have checked all this out before spending $30 billion to build wind turbines without the permission of the taxpayers who elected them.

At the present time electric cars run on lithium-ion batteries, which have proven to be quite flammable. A fire can start at any moment in one of these batteries. We have seen this happen many times. Manufacturers tried to solve this problem by using smaller cells with firewalls between them. They've tried to produce batteries with components that are less flammable and produce fewer harmful chemicals. Progress has been slow. It's difficult to extinguish an electric vehicle fire, which needs much more water to put out than a regular car fire. Fires can also reignite if the battery hasn't cooled sufficiently. It's important not to keep your electric car in a garage, since several houses have burned down when the electric cars stored in them caught fire.

"It's a serious issue that needs to be solved," said Dave Sargent, head of the automotive practice at research firm J.D. Power. "It's something the industry will have to get a grip on if EVs are going to be the future."[4] Any lithium-ion battery can catch fire, even cell phone batteries. Manufacturers have recalled some of their car models and are working hard to find a solution to this problem.

Cold weather can substantially reduce an electric vehicle's range. A study by Idaho National Labs revealed that an EV can lose 25% of its rated range when its heater is used in freezing conditions. This would also be true when the car's air conditioner is used. In that study it was also found that fast charging was 35% less efficient at freezing temperatures compared to room temperature.[5] 90% of the people who purchase an EV also own a fossil fuel powered vehicle that they use for long distance driving.

Instead of taking five minutes to refuel as with an internal combustion engine, it could take two to eight hours to charge an electric vehicle battery, depending on the charger. Some shopping centers and public parking lots now have electric vehicle chargers. You might be able to charge your car while shopping and running errands. Motel and restaurant owners are being told that providing a recharging service to customers will bring them more business. They are also told that they can apply for special government subsidies to lower their cost – again at the expense of the taxpayer.

courtesy of autoevolution.com

Legislation is being proposed in some states, such as North Carolina, to prohibit state or commercial entities from charging someone's vehicle for free unless they indicate on the receipt the amount of overall cost increase due to this "free" electrical charging. The government or commercial enterprise would also be required to provide an equivalent value of gasoline or diesel to those who don't drive EVs who are visiting the same establishment. And really, that is only fair since the resulting price increases at these restaurants or hotels are also borne by customers who don't own battery-powered vehicles.

Battery-powered cars average 3-4 miles per kWh of electricity. If you charge your car at home (as most people do), and drive 1,000 miles per month, at the price of 14.47 cents per kWh, you would pay $36.29 extra to the electric company. In California, with the average price of a kilowatt hour 22.81 cents, your cost would be $57.03 per month. That's with the 8-hour charger. "A level two EVSE requires special charging equipment and a separate, more powerful electrical circuit. You would also have to pay installation costs. On average, a level two EVSE system costs $400 to $6,500 for equipment, and $600 to $12,700 for installation."[6]A study by the Anderson Economic Group concluded that it is cheaper to operate a midsize fossil fuel powered car than a comparable EV.[7]

One of the worst aspects of battery-powered vehicle ownership is that it can be controlled from an external source. The government is requiring all car manufacturers to place a "kill switch" by 2026 in the car, but many already have these. Remotely activating this switch will make the car inoperable. This means that the owner of the car doesn't have ultimate control of it. Manufacturers are already charging consumers extra per month for heated seats, an option that is controlled from a remote location. Cars can also be set so that their speed doesn't exceed 65 mph or some other speed. We have to wonder why this was done.

Electric car manufacturers are raising vehicle prices because of the skyrocketing costs of some of the key materials needed for their batteries. The International Energy Agency reported that although battery costs had been declining for a decade, they are now expected to increase by 14% to $150 per kilowatt hour. One manufacturer blames the increased cost on the projected demand for minerals rising another 20% over the next four years. Korean auto manufacturers are considering a 30 to 40 percent increase in prices, blaming these price hikes on supply-chain disruptions related to COVID lock downs.

Governments are sinking tremendous subsidies into the purchase of these cars. Norway, where 65% of new car sales are battery powered, gives the EV buyer a $20,000 discount. An EV purchaser in the United States receives a $7,500 discount paid by taxes. Most battery-powered cars are bought by the richest 20% of the population. This is welfare for the rich at the expense of middle-income and poor people.

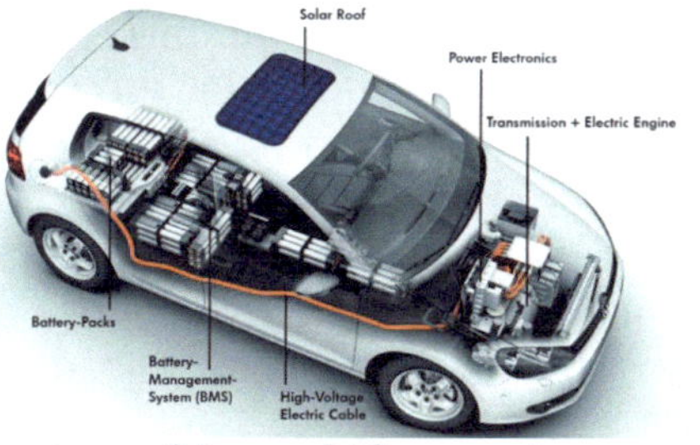

courtesy of downshiftmagazine.com

At this time (2023) the average price of a battery-powered car is $58,940, compared to $44,182 for a car powered by a carbon-based fuel. [8] With 60% of Americans living from paycheck to paycheck, most cannot afford to pay almost $60,000 for a vehicle, even though the government will reimburse EV buyers $7,500, paid for by the taxes of the people who can't afford battery-powered vehicles. The EPA has now released new tailpipe emission rules that are so strict that cars with internal combustion engines (ICE) won't be able to comply. This is how the government will force Americans to buy electric cars. It is estimated that by 2030 over half of all cars sold in the US will be battery powered. The federal government is also giving tens of billions of dollars of subsidies to battery producers and for the construction of millions of new electric vehicle charging stations.[9] And of course this is all being done without asking or even informing the public of what they are doing.

Between the number of nuclear and coal electrical generation plants going off line and not being replaced and the unreliability of wind and solar installations, we could be facing some really serious problems getting from one place to another if everyone is driving an electric vehicle. According to Mark Mills, the seven-fold increase in demand for minerals required for this will be the greatest ever in all of human history. There is already a shortage of copper, cadmium, nickel, cobalt, zinc, aluminum and rare earth metals. Since copper is also used for electrical cables, there will be a serious shortage of that metal in a year or two. It takes approximately sixteen years to establish a new mine.[10]

Over half of the world's cobalt currently comes from the Democratic Republic of the Congo, one of the poorest countries on Earth. China is directing many of these mining operations in Africa. Recently it was revealed that *thousands* of children in the Congo, some as young as four years of age, have been put to work in the arduous, dangerous and dirty process of mining for cobalt under horrendous conditions. Anyone who purchases a product with such a battery knowing that they are contributing to this atrocious practice is complicit in the abuse of these children. According to Bloomberg, "For consumers seeking to lower their carbon footprints, the environmental and social costs of electric vehicles may be greater than they realize." Many of these materials are produced in countries such as China and Iran, that are known for serious, widespread human rights violations.

Children Mining Cobalt in the Congo

Mining and processing the materials needed for EV batteries is an environmental disaster. "The production of *just one* EV battery requires 25 pounds of lithium, 60 pounds of nickel, 44 pounds of manganese, 30 pounds of cobalt, 200 pounds of copper, and 400 pounds of aluminum, steel, and plastic. You must process 25,000 pounds of brine for the lithium, 30,000 pounds of ore for the cobalt, 5,000 pounds of ore for the nickel, and 25,000 pounds of ore for copper. All told, about half a million pounds of the earth's crust is dug up *for just one battery*."[11] In Indonesia, where nickel for EV batteries is also produced, "polluted air and water are causing respiratory problems, sickness, and eye injuries and destroying forests and fisheries. The rush to expand production has pushed local communities and infrastructure to the brink of collapse."[12]

Lithium Mine - courtesy of survivethenews.com

Lithium mining is one of the most ecological destructive on the planet. At just one lithium mine 35-40 Caterpillar haul trucks use up almost 18 million gallons of diesel — a fossil fuel — per year. Two other minerals needed for lithium-ion batteries are nickel and cobalt (also used in smart phones, laptop computers and some power tools). The production cost of these minerals has multiplied several times over the last seven years. Because Russia is a major nickel exporter, its invasion of Ukraine is interfering with the production of nickel.

Battery-powered vehicles use much more aluminum than conventional vehicles. Ford's F-150 battery-powered truck contains 682 lbs of aluminum, mostly mined and refined in Brazil. There is a class action lawsuit going on there against Hydro, a Norwegian company. "The alumina made here is produced at the cost of a lot of misery," proclaimed Ismael Mores, a lawyer working on this lawsuit. Here bauxite [aluminum ore] is mined and converted to alumina and a toxic residue which leaches into rivers and creeks.

Mark Mills, a senior fellow at the Manhattan Institute and professor at Northwestern University, recently testified before Congress. He says that before an EV is sold, it has already consumed 20-30 barrels of petroleum equivalent. Two dozen minerals are needed in the manufacture of a battery-powered vehicle, which requires four times as many minerals as a traditional car.[13]

Lithium-ion batteries are usually replaced every seven years for smaller vehicles and three to four years for larger vehicles, such as vans and buses. Over a period of time, the number of miles a car can be driven between battery charges declines. The car owner must recharge the battery more and more frequently. The guarantee offered for such a battery is honored only if it wasn't damaged by an incidental problem, such as a collision or a stone flying up and damaging one of the 3,000 or so cells making up the battery. There have been cases where a relatively new battery had to be replaced because one or more of these 3000+ cells had been accidentally damaged.

Since batteries are full of toxic chemicals, they can't just be placed into landfills. Like solar panels, they require recycling, an intensive process. Recycling these batteries is difficult and complicated because of the sophisticated chemical procedures involved. If this is not properly done, the heavy metals involved will contaminate water and soil. Extracting lithium from a used battery is five times as expensive as the cost of mining it. So there is not much incentive to recycle batteries. Therefore, every EV purchaser should pay the cost of recycling the car's battery at the time of purchase so that he won't just abandon the spent battery at the end of its life.

Inventors are working feverishly to correct these problems and develop electricity storage systems that can store more power for longer periods of time. But this may take many years if indeed they can ever solve these problems. How much time will pass before this happens is unknown. Those in power should have had **all** of these problems worked out before putting the world's people in a no-win situation.

---

1. Michael Sivak & Brandon Schoettle, "Relative costs of driving electric and Gasoline Vehicles in the Individual U.S. States", Feb. 2018.
2. Editorial, "Investor's Business Daily", 11/7/17
3. by Anderson Economic Group
4. https://www.forbes.com/wheels/news/battery-car-fires/
5. https://www.cnet.com/roadshow/news/electric-vehicle-range-shopping-tips/
6. https://www.investopedia.com/how-much-does-it-cost-to-charge-an-electric-car-5186285
7. Jamie L. LaReau, "Which is more expensive: Charging an Electric Car or Fueling a Car with Gas?" Detroit Free Press, 10/21/21
8. Sean Hannity, Fox News, April 13, 2023
9. Editorial, Institute for Energy Research, 4/14/23
10. Mark Mills, op. cit.
11. Ronald Stein, "Is it Unethical to Buy a Vehicle Powered by a Lithium Battery?" 6/13/22 "Issues in Insights "
12. Ibid.
13. www.instituteforenergyresearch.org > renewable > magic-renewables-mark-mills-revisited

Chapter Ten

Sustainable Energy Alternatives

"Consensus is the last refuge of scoundrels." — Michael Crichton, Author

Nuclear energy is our best option.

Currently nuclear energy produces about 19% of electricity in the US. The United States currently has 55 commercial nuclear power plants, the average age of which is about 40 years.[1] Many of the older nuclear generators are being decommissioned and not replaced.

The nuclear meltdowns at Chernobyl in 1986 and Fukushima in 2011 made many people distrustful of nuclear power. As a result, very few nuclear generating plants have been built in the past thirty years. Many of the older uranium reactors are wearing out, and some are leaking. The good news is that the latest nuclear reactor designs are much safer and, in some cases, will be able to use much of the radioactive nuclear waste that's been sitting around. This will also make nuclear power more affordable.

In contrast, over 30,000 people in the US have died from black lung disease, a direct result of the coal industry, since 1968. Coal electricity generation also produces much more toxic waste than nuclear does. So for these reasons we need to reduce our dependence on coal for power whether or not it has anything to do with climate change. But at the present moment doing so would condemn millions of the world's poor to misery and death.

There have been three major nuclear accidents: Three Mile Island (1979), which killed no one, Chernobyl (1986), which killed about 50 people, and Fukushima (2011), which killed no one. Considering the amount of electricity generated worldwide by nuclear reactors, there have been far fewer casualties from nuclear power than with any other source of energy.

No one was killed or injured in the partial core meltdown at Three Mile Island in 1979, and it leaked no appreciable radiation. This plant's containment system worked.

Although the Fukushima accident was much more serious, no one was killed there from radiation exposure. This accident resulted from a tsunami that was caused by an undersea earthquake registering 9.0 on the Richter scale. The resulting series of waves penetrated about ten kilometers into the land, killing over twenty thousand people. The water leaped over the seawall, disabling the plant's backup generators and preventing them from cooling down the reactors. The explosions occurring in these reactors were not from the meltdowns. They were explosions of hydrogen that had accumulated inside of the reactors after it separated from oxygen in the extremely hot water vapor. So the containment structures did their job of holding in the nuclear radiation. Unfortunately, some radiation leaked into the Pacific Ocean.

The accident at Chernobyl was by far the most serious. It directly killed about fifty people. This core meltdown was due to a poor engineering design that used graphite instead of water as a moderator (used in most designs) and a foolish test carried out by the plant's operators.

"According to an authoritative twenty-year study by the International Atomic Energy Agency and the World Health Organization, over time this fallout will most likely be responsible for four thousand deaths among the surrounding population."[2] Yet, as bad as this is, considering the amount of energy that has been produced by nuclear energy, it is minor compared to the thousands of deaths caused by coal, petroleum and hydroelectric energy. It has been estimated that for every nuclear generating plant replacing a coal power plant, hundreds of lives would be saved.[3]

How Nuclear Energy Works

In the 1930s it was discovered that when a neutron is shot into the nucleus of certain unstable atoms such as uranium-235, each atomic nucleus splits into two smaller elements, e.g. barium and krypton. The isotopes (types of atoms) that can be split — U-235, U-233 and Pu-239 — are said to be fissile. Each split atomic nucleus releases a great deal of energy plus two or three additional neutrons, which are then free to smash into nearby atoms in a chain reaction. The products of this reaction weigh slightly less than the original material.

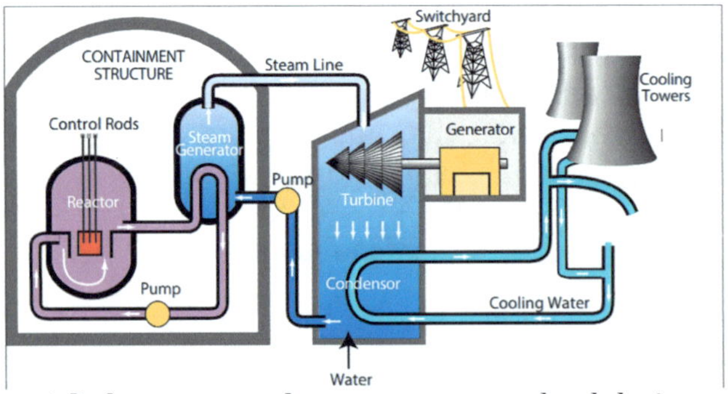

A light water nuclear reactor: standard design

The tiny amount of matter that was lost is converted into a tremendous amount of energy: almost a trillion times as much energy as the matter from which it was converted. This follows Einstein's equation $E = MC^2$. "E" is the energy produced, "M" is the matter (mass) being converted into energy, and "C" is the speed of light, 300,000 kilometers per second. When you square that large number (300,000 X 300,000), you get an unbelievably large number — almost a trillion. Nuclear energy is concentrated energy, producing about two million times as much energy per gram as the energy released from breaking chemical bonds in the combustion of coal or natural gas.

Just as in coal and natural gas plants, the heat energy produced in nuclear reactions is used to heat water to produce steam. Because steam occupies over a thousand times as much space as liquid water, it exerts a tremendous force that is used to spin a turbine and an electromagnet. This changing magnetic field produces electricity.

Over 99% of the uranium found in the earth's crust is U-238, a form of uranium that can't be used directly, because it's not fissile. It can't be split as U-235 can. In order to make fission work, some of this U-238 must be removed from the mix to raise the percentage of U-235 to about 3%. This process, called enrichment, is complicated and costly. Thus, the U-235 isotope is rare and expensive.

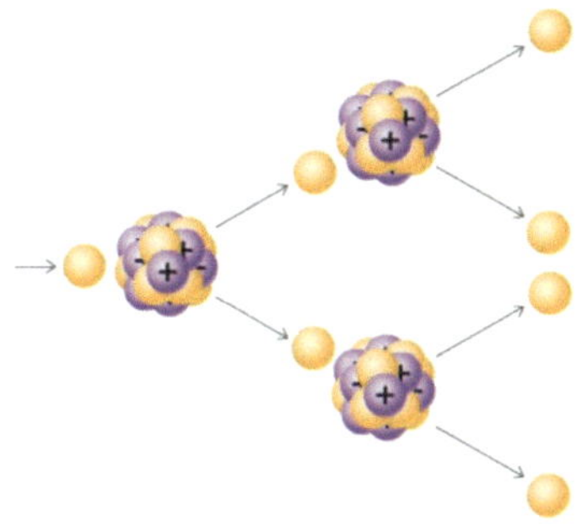

An unstable nucleus hit by a neutron breaks into two smaller nuclei.

In addition to the expense of enrichment is the problem of keeping the water used in traditional light water reactors at the pressure needed to keep it from turning into steam at 100°C. It's like a pressure cooker. If the pressure is accidentally released, it can explode. Water is the moderator, slowing down the neutrons so they can more easily make contact with the U-235 nucleus being split. If this water accidentally escapes as steam, it can no longer do its job of coolant and moderator. In that case, the reaction would be out of control and could possibly cause a meltdown. This is why light water reactors must be housed in huge, expensive containment buildings. In a traditional nuclear reactor, solid fuel pellets of uranium encased in zirconium are inserted into fuel rods.

After about 1% of this nuclear fuel is used up — perhaps being converted to the elements strontium and xenon — the pellets swell, cracking their protective coatings. At this point they are unusable and must be replaced, creating nuclear waste. Because it isn't fissile, an atom of U-238, which makes up about 97% of this enriched fuel mixture, doesn't split when hit by a neutron. But its nucleus can ***absorb*** the neutron, increasing its weight and changing it into another radioactive element, plutonium-239. Plutonium is fissile and also very toxic. This most dangerous of all elements, plutonium, is the material of which nuclear bombs are made. Plutonium can also be used to power a reactor. So traditional uranium reactors are a breeding ground for plutonium, which can be used either in nuclear reactors or bombs.

A totally different nuclear process which was successfully tested for about five years, from 1965 to 1969, uses thorium, another radioactive element. Because thorium — atomic number 90 — takes longer to decay than uranium, it isn't nearly as radioactive. The half life of uranium is 4.5 billion years. This means that it would take 4.5 billion years for half of a sample of uranium to change into another element. The half life of thorium is over 14 billion years. You can hold a lump of thorium in your hand and it won't hurt you.

When hit by a neutron, the nucleus of a thorium atom doesn't split, but just like uranium-238, it absorbs the neutron. This starts a series of reactions in which thorium is changed into another isotope of uranium, U-233, which is fissile. When hit by a neutron, U-233 splits into two smaller elements, just as U-235 does, releasing a huge amount of energy plus two more free neutrons that can smash into more U-233 atoms in a chain reaction.

Thorium is four times as abundant as uranium. Even better, it doesn't need the expensive enrichment process required by uranium. So 100% of the thorium that is mined, compared with 1% of uranium, can be used for fuel.

Because thorium is not used as a solid fuel but instead is dissolved in molten salt, 99% of it can be used in the reaction, compared with only 1% of a solid uranium fuel pellet that swells and cracks when its "daughter elements" are produced. *This means that thorium can produce electricity about a thousand times more inexpensively than uranium.*

Instead of water, the molten salt in which the thorium is dissolved is the coolant in this reaction. Salt has a much higher melting point than water. It melts at about 700°C - 1000°C. So this system doesn't need to be pressurized, as water does in a light water reactor. As a result, no containment structure is needed for this type of reactor. The power plant is thus much smaller and less expensive than a light water uranium reactor. In the event of a power failure, the safety system works automatically. Not requiring human intervention, it is foolproof.

An accident such as the one at the Fukushima Daiichi plant — where the backup diesel generators failed — could never happen with a thorium molten salt reactor. Beneath the system of pipes through which the molten salt circulates is a "plug" of salt kept solid by a fan blowing cold air on it. If the power should ever fail or is turned off, the fan will stop running, and this plug will melt, allowing the hot molten salt mixture to drain out into larger tanks below. At this point the reaction cools down and stops, because the atoms are now too far apart to support a chain reaction. This type of reactor would never experience a meltdown such as those which occurred at the Chernobyl, Fukushima and Three Mile Island nuclear plants.

Thorium is so abundant in the Earth's crust that we will never run out of it. A lump of thorium about the size of a tennis ball could provide all of the energy a person needs for a lifetime: for transportation, heating, cooking, light, etc. And all of this would cost only about $150 US (in today's dollars). Compared with today's cost of energy, this amounts to $1 - $2 per year (in today's dollars) to supply all of a person's energy requirements for a lifetime.

The low cost electricity this reaction would produce could be used to split water molecules to produce hydrogen gas economically. We could thus have hydrogen-fueled automobiles. They wouldn't need large batteries. When hydrogen burns, the only product is water vapor. The low cost power obtained from molten salt reactors could also be used in desalination plants to provide all the fresh water people need. Desalinated water could also transform Earth's deserts into fertile farmland. This is not a pipe dream.

Just one thorium mine could provide all of the world's energy needs for a year. Hundreds of thorium mines have already been discovered. Because thorium is a waste product of the mining of rare earth metals currently used in electronics and for the magnets used to power wind generators, there are currently piles of thorium sitting around near these mines.

By now you might be wondering why, if thorium has so many advantages over uranium, it isn't being used instead to produce electricity. There are a couple of answers to this question. First, the initial research leading to the development of nuclear power occurred during World War II. The leaders of the United States feared that the Nazis might construct a nuclear weapon.

At the conclusion of the war, the focus changed to peacetime uses of nuclear power. Uranium was the fuel of choice for leaders who anticipated the possibility of having to use a bomb in some future war. Some scientists believe that if nuclear technology had been developed during a time of peace, thorium, which cannot be used for bombs, would have been the nuclear fuel of choice.

A thorium reactor that ran for almost five years at Oak Ridge Labs in Tennessee was invented by Dr. Alvin Weinberg, the same man who invented the light water reactor that uses U-235. Believe it or not, when this brilliant man expressed his concern over the safety of the light water reactor, he was fired, and the thorium project was canceled.

Thorium power plant in the Netherlands

The nuclear accidents at Chernobyl and Fukushima, along with the realization that wind and solar power are not the answer to the world's energy needs, have inspired a renewed interest in thorium as a nuclear fuel. But some technical problems must be worked out before thorium on a commercial scale becomes a reality.

One problem is that the molten salts in which thorium is dissolved are corrosive at higher temperatures. An alloy containing mostly nickel to house the reactants has been developed. Another glitch is that the element protactinium, the intermediate product into which thorium changes before becoming uranium-233, tends to absorb some of the neutrons that are needed to keep the reaction going. So it must removed from the mix at a certain point in the reaction. These problems are being solved. Development of a working thorium molten salt reactor is underway in China, India, the United States and the Netherlands, where a working thorium molten salt reactor began experimental operation in 2017.

Nuclear fusion could be the answer to the world's energy needs.

Fusion is the nuclear reaction that takes place in the sun to produce just about all the energy that we receive on earth. Fusion produces even more energy per atom than fission. In a fusion reaction deuterium (H2) hydrogen atoms are squeezed together with intense pressure and heat to produce helium. Deuterium atoms are twice as heavy as regular hydrogen, because each of them also contains a neutron as well as a proton. The positively charged protons, repelling each other, must be put under tremendous pressure at extremely high temperatures to join together to make a helium atom. When they are pushed closely together, the nuclear strong force, pulling protons together at close range, is able to overcome the opposite force, electromagnetism, that is trying to push them apart at the same time. Inside every atom there is a war going on between these two forces.

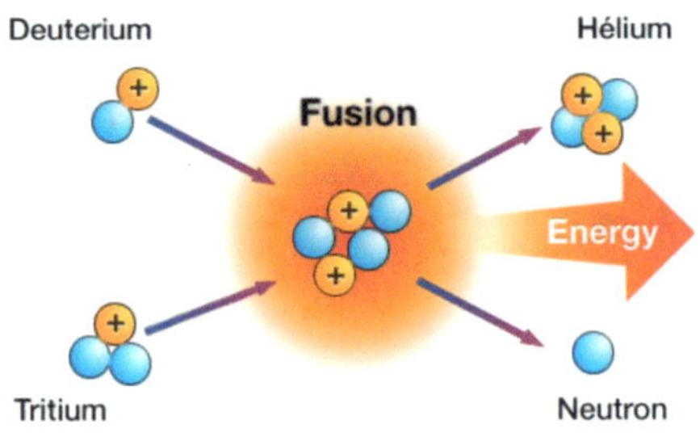

Because of the extreme heat and pressure required to do this, fusion is extremely difficult to accomplish. It is impossible to find a reactor shell that won't melt from the extremely high temperatures (millions of degrees) needed in this reaction. So the reactants must be held together by powerful magnetic fields. The energy produced when these atoms fuse must be greater than the energy needed to push them together. Otherwise the process wouldn't be practical for commercial use.

Recently there have been some promising developments in fusion technology. Experimenters at the Massachusetts Institute of Technology (MIT) tested a new type of reactor shaped like a sphere. Rather than trying to join hydrogen atoms together, this process would join hydrogen and boron to produce carbon. These reactants would be heated by lasers.

Meanwhile, in Germany another fusion reactor known as the Wendelstein 7-X Stellerator reactor is being readied for operation. This project is reportedly a long way from being able to produce sustainable energy. At Texas A & M University, a team of researchers experimenting with helium at high temperatures are also finding the task frustrating. When helium heats up to these high temperatures, it begins to bore holes resembling veins in the metal casing.

Numerous other groups continue to pursue the nuclear fusion dream. The largest fusion experiment, located in southern France, is called Iter Tokamak. Thirty-five nations are participating in the construction of this massive dough-nut-shaped fusion generator.

Iter Tokamak

Iter Tokamak is by far the most complicated machine ever built as well as the most complex scientific endeavor in human history.[4] To hold the extremely hot hydrogen and helium atoms together at 150 million degrees, large superconducting magnets must be used. Strangely, in order for the magnets to be superconductive, their temperature must be held at a frigid -269°C. So some parts of the machine are exceptionally cold, while others are exceptionally hot. Despite sixty years and billions of dollars spent on this research, producing fusion on a commercial scale could still be decades away.

Hydroelectric and Geothermal Power

Hydroelectric power uses the downhill force of running water to spin the blades of a turbine that turns an electromagnet. The changing magnetic field created by the rotating north and south magnetic poles is converted to electrical energy. Hydroelectric power is 90% efficient, making it the most efficient energy source available. But the initial construction and startup costs of a hydroelectric power plant can exceed a billion dollars. So it's difficult for third-world nations to get the funds to build hydroelectric projects. “In 2021, hydroelectricity accounted for about 6.3% of total U.S. utility-scale electricity generation and 31.5% of utility-scale renewable electricity generation.”

Grand Coulee Dam

Yet hydroelectric power plants have their downside. They can cause great environmental damage by flooding large land areas on which natural plants and animals exist as well as converting river valleys once used for agriculture into virtual wastelands.[5] The presence of a new dam can even affect the migration patterns of fish. Hydroelectric dams can be dangerous and life threatening. The worst disaster in the production of electrical energy occurred in 1975 when a dam built in China twenty years earlier was unable to accommodate a huge amount of water from a typhoon that had dumped over 40” of rainfall on the area within 24 hours. The sluice gate couldn’t take the water fast enough. A huge wave crested over the dam and rushed downstream at 31 mph to create an instant flood. Approximately 26,000 people died from this flood, and another 145,000 died later from famine and disease. Almost six million buildings collapsed in this disaster. The risk from nuclear plant meltdowns pales by comparison to this disaster.

Geothermal energy is another sustainable alternative to the combustion of carbon based fuels. Heat deep in the earth’s crust is generated partly by the decay of radioactive minerals in the lower crust. Some parts of the water table under Earth's surface are heated by hot rocks close to concentrations of hot melted rock, called magma, that has been pushed up from the earth's mantle. Sometimes this water is even superheated to steam, which can be used to spin turbines connected to electrical generators.

One such operation in the US is a plant north of San Francisco called the Geysers. This plant produces almost one GW (billion watts) of electrical power.

Over thirty nations are producing geothermal energy. Geothermal heat left over from electrical generation can also be used for heating buildings. Iceland, where there are many volcanoes, uses geothermal heat for heating huge greenhouses, in addition to electrical power generation.

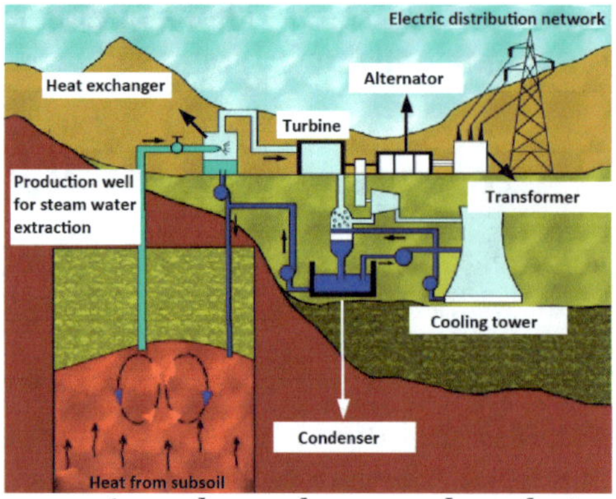

A geothermal power plant design

But there are also problems in extracting heat from the Earth in this manner. A Swiss newspaper article dated December 10, 2009, describing a geothermal project in Switzerland, states: "The authorities in canton Basel City say they will cancel a geothermal energy project, which three years ago caused minor tremors that damaged many buildings. A risk analysis study published on Thursday found that the danger of setting off more earthquakes was too great if drilling at the site resumed."

Nevertheless, as drilling methods improve, many more such projects are being planned – and not just for places where there are natural hot springs. In southern Saskatchewan, near the US border, a series of wells were drilled 3.7 km (2.3 miles) vertically and connected by drilling horizontally. Water run through this system picks up heat and moves back up to ground level by convection. The resulting steam drives a turbine that generates 20 MW (million watts) of electricity, enough for 20,000 homes.

Drilling technology has improved by a factor of 10 in the last ten years. Many people now working in oil fields, who understand the technology of drilling, can use their skills for geothermal operations. So we are probably going to see many more geothermal energy plants built in the near future.

---

1. Zubrin, op cit., p. 133.
2. Ibid.
3. Ibid., p. 142
4. https://www.theguardian.com/environment/2017/dec/06/iter-nuclear-fusion-project-reaches-key-halfway-milestone
5. https://www.eia.gov/energyexplained/hydropower/

Chapter Eleven

The Final Analysis

> "If you tell a lie big enough and keep repeating it, people will believe it. The lie can be maintained only for such time as the state can shield the people from the political, economic and/or military consequences of the lie. It thus becomes vitally important for the State to use all of its powers to repress dissent, for the truth is the mortal enemy of the lie, and thus by extension, the truth is the greatest enemy of the State." — Joseph Goebbels, Third Reich Minister of Propaganda

Carefully studying and verifying all of these facts, one can only conclude that the climate change hypothesis doesn't stand up to scientific scrutiny. It is a political issue (see Chapter 6). To capture the moral high ground, a group of the world's elite with plans to establish a totalitarian world government have usurped environmentalism, an important cause. At the same time the way they live their lives demonstrates that they themselves don't care about the environment. John Kerry, the US climate czar, defended his flying all over the world in a private jet by calling it, "The only choice for somebody like me who is traveling the world to win this battle."[1]Private jets emit up to forty times as much carbon dioxide per mile as traveling on an airliner.

These people have clearly stated that they want to reduce the world's population from the current 7.5 billion to as low as 500 million (a 15-fold reduction). In 2010 Bill Gates made it very clear in a Ted talk that he thinks the world's population should decrease to near zero. To hear this, you can access a one-minute segment of Gates' speech by going to: "Bill Gates – Population – Vaccines" YouTube channel: Panagiotis Liakos.

As explained in Chapter 6, the elites founded the Club of Rome in 1968. "The First Global Revolution", a report they wrote in 1991, gave their strategy away in this statement, **"In searching for a new enemy to unite us, we came up with the idea that pollution, *the threat of global warming*, water shortages, famine and the like would fit the bill."**

The drastic measures outlined in the Green New Deal, [2] which the Biden regime is now putting into effect in the US, call for completely replacing by 2030 the combustion of "fossil fuels" (coal, natural gas, gasoline and diesel) with "renewable" sources: wind, solar and biofuels. They aim to replace all gasoline and diesel engines with battery-powered cars, trucks, farm tractors and construction vehicles. In this far-fetched scheme all electric utilities would rely on wind, solar and biofuel-powered generators. Yet their attack on the Nordstream Pipeline released the greatest amount of greenhouse gases in human history.

From 2010 to 2018 the world spent $3.7 trillion (US) promoting the increased use of "renewable" wind, solar and biomass. However, in 2020 these three energy sources were providing only 5% of the world's energy. Hydrocarbons still provide 84% of the energy we use.[3] Yet China emits the majority of the world's CO_2 emissions.

The severe winter storm that hit Texas and nearby areas in the middle of February, 2021, demonstrated that not only is the climate not warming but that nearly half of their wind turbines, which had cost Texas residents upwards of $30 billion, froze up and ceased operating. As a result, millions of people had no electricity for several days. Over 200 people died from lack of heat.

Texas wind turbines

The media blamed this situation on problems with natural gas, carefully leaving out the fact that the wind turbines which had stopped operating had put more pressure on the natural gas plants than they could handle. And of course, climate change was blamed for the cold weather.[4] "Scientists who study climate change claim that Texas' extreme weather is due to the heating of the Arctic."[5]

The United Nations Intergovernmental Panel on Climate Change (IPCC) has announced that humans are responsible for the small amount of warming that has occurred since the end of the Little Ice Age at the end of the 1700s, totally ignoring natural causes of climate change. *Without empirical evidence* to back up this hypothesis they claim that if their dictates are not followed, the earth's average temperature will increase to dangerous levels in the near future. This, they say — also without any evidence — will cause the extinction of many species of plants and animals, extreme weather events and danger to food supplies, human health and economic growth. The IPCC estimate of the cost to remedy this is $2.4 trillion US per year.[6]

Yet many prominent scientists — including Nobel laureates — have expressed serious doubts about these claims. The number of skeptics is growing each year that "tipping points" come and go and are changed to later dates. Claiming that the issue is settled, the doomsayers refuse to debate the issue with other scientists. Instead they call those who question them "climate deniers" or even "science deniers". And now they are even censoring voices of disagreement.

The spokesman for human-caused climate change appears to be TV personality and clown, Bill Nye, the Science Guy (who isn't a scientist). Nye along with some politicians has even called for jailing climate skeptics under RICO, the Racketeer Influenced Corrupt Organizations Act.

When Tucker Carlson asked him in 2017, "To what extent has the rate of climate change been increased by human activities and how much of it was caused by natural causes?" Nye was unable to provide an answer. If the science is indeed settled, the "experts" should certainly know the answer to this important question. “In spite of all the well-publicized concern about global warming, you must understand that there is still considerable uncertainty among scientific experts about a number of critical factors which determine global warming,” NOAA administrator John Knauss admitted[7]

Richard Lindzen, PhD, is one of the world's most respected climate scientists. As the Alfred P. Sloan Professor Emeritus of Meteorology at MIT, Dr. Lindzen has made important contributions to climate science. Here is what Dr. Lindzen had to say about climate alarmism:

> "What historians will wonder about in future centuries is how deeply flawed logic, obscured by shrewd and unrelenting propaganda, actually enabled a coalition of powerful special interests to convince nearly everyone in the world that CO_2 from human industry was a dangerous, planet-destroying toxin. It will be remembered as the greatest mass delusion in the history of the world – that CO_2 the life of plants,was considered for a time to be a deadly poison."

In the US the Biden administration is now enacting “Green New Deal” measures *without the consent of Congress*. If the Green New Deal is implemented, your life will never be the same again. It will involve a complete reshuffling of government and society. Your taxes will skyrocket. You will be told what kinds of food to eat. Meat eating won’t be permitted, because the plan’s proponents claim that meat production produces methane, a greenhouse gas existing at only 2 parts per million. The raising of cattle and other farm animals for meat will be prohibited. The excuse for this is that livestock emit methane (CH_4 or natural gas). They never tell you that methane exists in a very tiny amount: 2 parts per million (0.0002 percent). The amount of methane emitted by these farms would have to be multiplied by many times, to ever be considered a threat. By subjecting farmers to a multitude of unreasonable regulations, the world's food supply has been placed in great jeopardy.

Unless you own an electric car, you’ll be traveling on railroad lines, walking or riding your bike. The proponents of this plan also want to phase out air travel. Under the Green New Deal every building in the United States would be re-engineered for energy efficiency. Every aspect of your life would be micromanaged by unelected bureaucrats. This is true socialism at best, but most likely, totalitarian communism, even though those terms are omitted from their documents.

Please note that all of these measures will apply to us peons, not to the elites who are pushing for them. They will still have their multi-million dollar mansions and continue traveling in their private jets to climate conferences around the world.

And, as said before, this is really a political issue. Alexandria Cortez's (AOC's) former chief of staff, Sycat Chakrabarti, said in a recent meeting, "The interesting thing about the Green New Deal is it wasn't originally a climate thing at all. Do you guys think of it as a climate thing? Because we really think of it as a how-you-change-the entire-economy thing."[8]

The Secretary General of the United Nations, Antonio Guterres, a self-proclaimed socialist, stated, "We need to make sure we reverse the trends. It's now clear we are coming to a point of no return. To spend those trillions of dollars and not use this occasion to reverse the trends and massively invest in the Green economy will be an unforgivable lost opportunity."[9]The Rockefeller Foundation announced their partnership with the World Health Organization (WHO) by contributing five million dollars "to cultivate global networks for pathogen detection and strengthen pandemic preparedness capabilities, including broadening surveillance for diseases worsened by rising temperatures and extreme weather." [10]One key word here is "surveillance". You can draw your own conclusions.

To help pay for climate mitigation, the world's central banks have printed and put into circulation trillions of dollars. It is a basic fact of economics that putting more money into circulation creates serious inflation, devastating our savings and retirement accounts. Inflation is the hidden tax. We need only to look at what recently happened in Venezuela. The rate of inflation that just occurred there under a socialist dictatorship was as high as 1.6 million percent. Before allowing our lives to be taken over in an experiment to "prevent the world from ending", we need to ask, "Is the earth really headed for disaster if we fail to go through with these drastic measures which will shake our way of life to its very foundation?"

As a voter or policy maker, you need to know as much as possible about this issue. This way you'll know which candidates for office to support: those who believe we can change the climate by reducing CO_2 or those who believe that whatever climate change is happening is not a crisis.

As stated before, the entire "climate change" scenario that was cast upon the world in the 1970s — first as a global cooling crisis and then as a global warming crisis — really takes advantage of the world's need for energy, without which very little can be accomplished. In the not-too-distant past energy was provided mostly by animals and by human slaves as well as by burning wood. Wood, as well as coal, was used in the first steam engines at the beginning of the Industrial Revolution in the early 1800s. Petroleum was discovered in 1859.

As the 21st century progresses, the world's people will need more and more energy, on which all life depends. The world currently relies heavily on carbon-based fuels, which still provide over 84% of the world's energy. It can't be denied that there are many health hazards associated with coal mining. But there are drawbacks to every form of energy production. Alex Epstein noted, "When evaluating any issue, we must look at the full context, carefully weighing the benefits and side effects of every option." Our shortsighted politicians, who obviously have never studied the issue, have utterly failed to do this.

This oil well, from which 100% organic material is pumped out of the ground, requires about 500 to 1,000 square feet. The petroleum is then transported safely through pipelines to a refinery to be processed into usable compounds that are made into over 6,000 products for society. It is the basis of modern transportation, including aviation, commercial ships, and the military.[11]

It would be impractical and inhumane to hold back from the world this inexpensive and dependable source of energy that people desperately need **now** for heat, transportation, agriculture and food storage. For those in power to deny the world's people access to carbon based fuels without offering a viable and sustainable alternative, such as nuclear energy, would wreck the entire world economy. But this is what the founder of the UN IPCC, Maurice Strong, said he wanted to do when he asked, ***"Isn't the only hope for the planet that the industrialized civilizations collapse? Isn't it our responsibility to bring that about?"*** Withholding these energy sources from people would also sentence millions to death from starvation, freezing and disease.

Most everyone agrees that to reduce pollution, cleaner methods of producing energy other than burning hydrocarbons need to be found. But the fact remains that ***at the present time*** the world's people rely almost exclusively on these dependable, low-cost sources of energy that have enabled humanity to thrive. No one can deny that affordable, abundant energy from coal, natural gas and petroleum has made life for billions of people much better than it was when all our energy came from burning wood, human slaves and beasts of burden for farming, heating and transportation.

As Christopher Monckton so aptly stated, "Efforts to make a nonexistent problem go away would hurt the poor the hardest."[12] It is immoral to forbid third-world nations the same opportunities that boosted the richer nations to their current level of affluence. Energy now provided by coal and petroleum could be gradually replaced with safe nuclear reactors.

If human-caused CO_2 emissions — responsible for less than 4% of the slight increase in the trace gas carbon dioxide — are not significantly warming the earth, then it follows that the trillions of dollars already spent and planned to reduce these emissions will make little or no difference in the resulting temperature. This money will have been totally wasted.

The Small Business Administration has estimated that compliance with all of the regulations, many of which are state regulations already in place, is already costing US taxpayers over $1.75 trillion per year — about 12%-14% of the US Gross Domestic Product (GDP) and about half of the $3.5 trillion per year that the federal government currently spends.

This includes huge subsidies for the wind, solar and biofuel industries, administration, propaganda campaigns, and tax breaks [welfare] for the affluent who install solar panels on their homes and buy electric vehicles. These wasted trillions of dollars could have been spent on solving the earth's ***real*** environmental problems, such as recycling plastics, metals and paper, curbing deforestation, and removing pesticides, dioxin and nitrates from our water supplies, to name just a few. Just the $3.4 billion spent each year to pump carbon dioxide into the ground [sequestration] alone could be used to provide clean water each year to a hundred million people who don't now have it, saving millions of lives.[13]

Several of the world's leading economists contend that the funds now being wasted could be much better invested in research and development of cost-effective, sustainable energy alternatives. When these resulting innovations come about, as they ultimately will, they'll quickly be put in place by the free enterprise system as has happened throughout history. Kerosene replaced whale oil in the 1860s. Automobiles replaced horses in the early 1900s, and the catalytic converter solved the air pollution problem posed by automobiles. The Green Revolution in agriculture saved millions from starvation in the 1970s.[14] Now many people are becoming aware of the almost unlimited potential available from nuclear energy that is awaiting civilization. These improvements were made under capitalism and free-enterprise. But improvements have *never* been made under communism.

Coal will continue to dominate the world's sources of energy until adequate and safe nuclear energy — either fission of fusion — becomes commercially available.

In contrast, wind turbines work only when the wind is blowing just right. They must be backed up 5/6 of the time by coal, natural gas or nuclear power plants. The same is true of solar voltaic panels, which don't work at night, in the winter or in cloudy weather.

In the 1970s we were told that in another twenty years the world would run out of petroleum. Since that time more and more oil deposits have been discovered. There are now about 65,000 known oil fields, many of which are gigantic. Over 900 of these contain at least five hundred million barrels (about 79,000,000 cubic meters) each. The oil companies don't want the public to know just how much petroleum there is and how cheaply it can be pumped out of the ground. Nevertheless, most people would agree that we shouldn't continue to burn coal when cleaner, cost-effective and sustainable alternatives, such as nuclear and geothermal energy are readily available.

It has become increasingly obvious that the three "renewables" — wind, solar and biofuels — being pushed on us by the climate alarmist cabal — have a limited application and can never replace carbon based fuels. Trying to force these on people with subsidies, cap-and-trade schemes, regressive taxes and even more draconian tactics will destroy the world's economies and end up killing millions of people. Further evidence that these "renewables" are ineffective substitutes for carbon based fuels is the fact that for several decades these alternatives have been subsidized by government grants – some of which go back to the 1970s. Without such subsidies these industries would have failed long ago.

Since the institution of "green" taxes the cost of heating in Great Britain has risen 63%. A million elderly British have been struggling to keep warm, and each year over 2,700 of these people have died from the stress of extended periods of cold weather. "There are hundreds of thousands of people who are struggling at the margin of survival...Over 15% of the population have an income below the poverty line...If [an increased] carbon tax comes in, these are the people that are most vulnerable," noted Dr. Sean Healy, head of Social Justice Ireland.[15]

Neil Chatterjee, the former chairman of the Federal Energy Regulatory Commission, noted, "We have prioritized ["Green Energy"] over reliability, and sadly it's going to take catastrophic events to recalibrate things."

In the last few years activists have put more and more pressure on politicians and the public with carefully planned demonstrations. Activists working with Extinction Rebellion glue themselves to statues and throw tomato soup at priceless paintings in art galleries. Across Europe students occupy schools and universities, shutting them down in month-long campaigns. Another activist group called Tyre Extinguishers claims to have disabled over 11,000 SUVs by removing the air from their tires. In their recruitment of climate warriors, they claim to have no leaders, an example of 5th-generation warfare tactics.

Extinction Rebellion Demonstration, courtesy of Breitbart.com

An organization called the Climate Emergency Fund says their mission is to "raise funds for and make grants to the disruptive, nonviolent climate movement." So far they have given away at least $5.1 million to forty four climate activist groups. The *New York Times* stated that people who have inherited oil company fortunes are actually contributing to this organization. One reportedly contributing this money is Aileen Getty, heiress of the Getty fortune. If there was really a climate emergency, these demonstrations wouldn't be necessary.

In 2012 a $25/ton carbon dioxide tax was introduced in Australia. This ten-year plan would have cost $162 billion. The goal was a modest 5% reduction in carbon dioxide emissions. This caused such extreme economic deprivation that the number of bankruptcies skyrocketed. Because Australia's CO_2 emissions amount to only 1.2% of mankind's total CO_2 emissions, this reduction would have made a difference in the earth's average temperature of only 0.00005°C after ten years.

If we extend those figures out to worldwide proportions, we find that the cost of reducing the earth's average temperature by even one degree Celsius would be $3.2 quadrillion ($3,200 trillion), an astronomical figure. This is assuming that the IPCC's estimates are correct. Reducing the temperature just 1/7 of a degree would thus cost $450 trillion. This cost, spread out equally among the Earth's 7.5 billion human residents, would amount to $65,000 for each man, woman and child.[16] And that's for just 1/7 of a degree. No one would even notice that. Would the sacrifice really be it worth it?

The alarmists tell us that it is our duty to make these sacrifices for future generations: our children and grandchildren. But if we bankrupt ourselves and wreck our economy, what sort of legacy are we really leaving them? It would make far more economic sense to simply adjust to any slight warming that might occur if indeed we are influencing the climate at all. In the meantime we need to develop advanced nuclear energy that produces hardly any toxic waste or other residue compared with fossil fuels. Advanced nuclear technologies could use up most of the current nuclear waste.

"Instead of implementing economically destructive policies of more taxes, regulations, and subsidies, federal and state policymakers should remove government-imposed barriers to energy innovation. Allowing all forms of energy to compete equally in a free market will enable the U.S. to make tremendous strides both for the economy and for the environment."[17]

When push comes to shove, since sane people would never volunteer to give up their precious liberty and the wealth achieved through the free enterprise system, only a fascist or communist totalitarian government backed up with big guns will ever be able to achieve such a feat. Don't think for one minute that this couldn't happen. It has been said that those who haven't learned from the past are doomed to repeat it. We only have to remember how the Nazi government in Germany controlled its people with similar lies that led to war and the deaths of millions. The same tactics were used in power grabs by twentieth-century communist governments, which were ultimately responsible for the deaths of over a hundred million people. We need to learn from history and also teach these facts to our children before it is too late.

In order to persuade people to do things that are repugnant to them — like accepting a worse standard of living and giving up their freedom — a lot of fear and guilt must be heaped on them. Most people recoil at the thought of war. So throughout history corrupt leaders have softened up the people's resolve to avoid war by using corrupt, state-run media to manufacture a barrage of false reports about the "enemy".

On Nov. 5, 2019, an article in Bloomberg reported that 11,000 scientists: so-called "experts", say we're having a climate emergency and the world needs fewer people. In this case the enemy is "climate change," and we are told that we're in a war.[18]And it is true that we are in a war: an information war. The best thing we can do is to learn as much as possible about this issue and be ready to ask questions of policy makers — your representatives and other government officials — and people you know who need to learn this information. We have provided some useful information in the appendices of this book that will help.

If you still have any doubts about whether We the People should follow the dictates of globalist elites, such as the World Economic Forum (WEF), consider the fate of several nations that tried to do so. The companies that belong to the WEF are committed to following its energy policy, abbreviated ESG. (Environmental, Social and Governance.) The unelected leaders of the World Economic Forum has set itself up as world leaders. They've gotten away with this, because the world's largest corporations have banded together under the WEF banner. ESG requirements are a set of standards for a company's operations that "socially conscious" investors [banks] use to screen potential investments. It is now clear that ESG investing produces inferior returns quite aside from the fact that the investment criteria make no sense to begin with.

Nations compete for high ESG scores to obtain loans from these banks.[19]The first country to be ruined by this insane policy is Sri Lanka, a small island nation located off the coast of India. In 2019 Sri Lanka was prospering. But in order to obtain a high ESG score, their leadership was sold on ideas published in a WEF article by Joseph Stiglets that if they would stop using "fossil fuels" and artificial fertilizer, they'd be rich by 2025.[20] Instead of this, the opposite happened. They've run out of both fuel and fertilizer, and because their crops were ruined, the starving people have revolted against their government. Even the *Wall Street Journal* called the country's failed "Green New Deal" a "human disaster" that decimated its economy and monetary system. Rajapaksa, the prime minister, was forced to resign by email as he was fleeing from the angry mob. This opened up a power vacuum that allowed the WEF to quickly put one of their own people into a position of leadership there.

A similar plight has befallen the country of Ghana, located on the west coast of Africa. Several years ago Ghana was doing well with one of the fastest-growing economies of the world. In 2015 this headline was published: "World Bank Approves Largest Ever Guarantees for Ghana's Energy Transformation." The World Bank promised Ghana, "technical assistance for energy sector reforms and the drafting of a new renewable energy law." As a result of this "guidance" by people in NGO's posing as experts who have probably had little, if any farming experience, Ghana has encountered a complete shutdown of its electricity and (in some places) water. They are also facing serious food shortages.

The government of the Netherlands (with an ESG score of 98.7) is forcing its farmers to get rid of at least 30% of their livestock (such as sheep and dairy cows). The farmers say that this action will put some of them out of business, with the result that the government will confiscate about half of their land, which has been held in these farm families for several generations. This has led to protests and riots. Farmers in Italy, Spain, Germany and Poland are also rising up in support of these farmers, knowing that they will probably be next. When interviewed by Epoch TV in 2022, Thierry Baudet, a member of the Dutch parliament, stated, "We have eight years [Agenda 2030] to fight the Globalist takeover...This is the single most existential fight in the history of civilization."[21]

The country of South Africa has been facing severe food shortages for at least a year, with 3.8 million people undernourished, according to the UN. South Africa has an ESG score of 91. They switched to "renewables" in 2015. So the same story has been repeated here and will also happen in the United States and other countries, like Canada and Australia, if people don't wake up to what has been happening in other parts of the world. We all must refuse to comply with these ridiculous requirements being put on us by people who obviously have no idea of what they're doing. Sadly, because the globalists control the media, these issues have received hardly any significant coverage.

The Green New Deal in the United States is being implemented without the consent of Congress or the people. The US Department of Energy is now warning that the entire US power grid is at risk of system-wide power outages, beginning in the summer of 2023. This is expected, because the Biden regime is taking fossil fuel power plants offline without allowing replacements. "The Biden administration is announcing a climate rule that would require most fossil fuel power plants to slash their greenhouse gas pollution 90 percent between 2035 and 2040 — or shut down."[22] They are literally cutting gigawatts of power. But, at the same time, due to the curtailing of gas appliances, more electric stoves and hot water heaters will be placed on the grid. This will strain it to the breaking point. [23]They are playing a political game with one of America's most critical security assets, the Strategic Petroleum Reserve. Biden is gambling with national security to achieve a few political points.[24]

The World Economic Forum and their operatives are now withholding the fertilizer that's absolutely necessary for farmers to produce food for the world. Without fertilizer using nitrogen made with natural gas, agriculture would support only about 1-5 billion people. Withholding fertilizer will lead to the genocide of **billions** of people. Massive human sacrifices are required to meet the CO_2 stabilization goals of the Kyoto Protocol, which requires an immediate 60% - 80% reduction in carbon dioxide.

When you realize that global warming alarmism is not based on observation and measurement, needed for a true scientific investigation, but on pseudoscientific computer models and countless lies, you will want to share these facts with others. Some say there is a war going on, involving everyone whether they know it or not. This is being called "Fifth-Generation" warfare in which psychological operations (psyops), designed to influence the perceptions and attitudes of people worldwide, are being implemented. It's time to call them out. You can choose to take action to fight back or roll over and allow yourself and others to be victimized. In another generation or so people will probably look back with wonderment on this strange period of history.

NOTE TO THE READER:

We hope this book has provided tremendous insight and value to you.

If you benefited by reading it, it would be greatly appreciated if you left a short review to let others know how they can also benefit by reading it.

Feedback also helps us learn what is most important to our readers. We welcome your comments and would love to hear from you.

Thanks,

Lynne

Faraday Science Institute (FaradayScience.org)

Email address: contact@faradayscience.org

---

1. Sam Dorman, "Kerry family private jet has taken 16 flights this year alone, records show", Fox News, August 5, 2021
2. https://www.congress.gov/bill/116th-congress/house-resolution/109/text
3. Robert Lyman, Friends of Science, April 2022
4. https://www.matchcollegiate.org/2021/04/25/the-texas-ice-storm-of-2021/
5. Ibid.

6. F. William Engdahl, "The Dark Story Behind Global Warming aka Climate Change", Oct. 16, 2018, journalneo.org

7. Tony Heller, RealClimateScience, 5/14/2023

8. Jack Crowe, "AOC's Chief of Staff Admits The Green New Deal is Not About Climate Change", Yahoo News, July 12, 2019.

9. Oliver Milman, "Antonio Guterres on the Climate Crisis: 'We Are Coming to a Point of No Return,' " The Guardian, June 11, 2021.

10. Press Release, "The Rockefeller Foundation and World Health Organization Announce Partnership to Expand Global Pandemic Preparedness in Era of Climate Change," TheRockefellerFoundation.org, May 23, 2023

11. Ronald Stein, "Is It Unethical To Buy A Vehicle Powered By A Lithium Battery?", Issuesinsights.com, June 13, 2022

12. Christopher Monckton, "Kicking the Poor in the Teeth: the Cruel Cost of Needless Mitigation," speech at the Heartland Institute, July 2, 2015

13. Joanne Nova, The Skeptic's Handbook, 2009

14. Ibid.

15. "Carbon Tax To Hit Lowest Earners Hardest, Says Social Justice Ireland,"https://www.thegwpf. om/carbon-tax-to-hit-lowest-earners-hardest-says-social-justice-ireland/, Jan. 4, 2019

16. Address by Lord Christopher Monckton, "Carbon Tax - Lord Monckton - Part 2," Jan. 5, 2012, YouTube video

17. https://www.heritage.org/energy-economics/report/assessing-the-costs-and-benefits-the-green-new-deals-energy-policies

18. Eric Roston, "Earth Needs Fewer People to Beat the Climate Crisis, Scientists Say," Bloomberg, Nov. 5, 2019

19. https://www.future.enverus.com

20. Tucker Carlson, "Tucker Carlson Tonight," Fox News, July 12, 2022

21. Roman Balmakov, "Facts Matter," Epoch TV, July 13, 2022

22. Alex Guillen, https://www.politico.com/news/2023/05/11/biden-rule-tells-power-plants-cut-climate-pollution-00095827, 5/11/23

23. Mike Adams, "Blackouts Coming", Brighteon Broadcast News, May 8, 2023

24. Jim Rickards, June 1, 2023

Appendix I

GLOSSARY OF TERMS

AGGI annual greenhouse gas index

AGW anthropogenic, or human-caused, global warming

albedo effect the fraction of the total light striking a surface that is reflected, depending on the brightness of the surface. Clouds and ice have a high albedo, because they reflect sunlight.

biofuel fuel (such as wood or ethanol) composed of or produced from biological raw materials.

CCUS carbon capture and storage

climate weather, usually in a certain area, over an extended period of time, 20 years or more

climatism the belief that humans are causing dangerous global warming

COP Conference of the Parties. The United Nations Climate Change Conferences are yearly conferences held in the framework of the United Nations Framework Convention on Climate Change (UNFCCC). They serve as the formal meeting of the UNFCCC Parties (Conference of the Parties, COP) to assess progress in dealing with climate change, and beginning in the mid-1990s, to negotiate the Kyoto Protocol to establish legally binding obligations for developed countries to reduce their greenhouse gas emissions.

cosmic rays highly energetic particles moving at nearly the speed of light that enter Earth's atmosphere

CRU Climate Research Unit, University of East Anglia in Britain

EPA Environmental Protection Agency

feedback processes that amplify (positive feedback) or diminish (negative feedback) forcing from natural or human causes

fission process of splitting the nucleus of an atom of a radioactive element to produce smaller atoms, releasing a great amount of energy

forcing (climate) direct warming influence on the climate

fusion process of forcing atoms of lighter elements to fuse together, using extreme heat and pressure, resulting in the release of a tremendous amount of energy

Gigawatt one billion watts

GISS Goddard Institute for Space Studies, part of NASA

HadSST3 a dataset that provides monthly SST anomalies on a 5°x5° grid for 1850-present. The anomalies are derived from a 30-year climatology spanning 1961-90. HadSST3 has been replaced with HadSST4.

hypothesis an idea or explanation that has not yet been verified by experiment or observation

IER Institute for Energy Research

IEA International Energy Agency; IEA is at the heart of the global energy dialog.

IPCC Intergovernmental Panel on Climate Change, a United Nations agency

isostasy maintaining equilibrium in the earth's crust such that the forces tending to elevate landmasses balance the forces tending to push landmasses down.

Kwh kilowatt hour; a unit of power, one thousand watts used for one hour A kilowatt of energy is the amount of power used by the average microwave oven.

Kyoto Protocol (1997) an international treaty extending the 1992 United Nations Framework Convention on Climate Change It commits state parties to reduce greenhouse gas emissions.

Mwh megawatt hour; one million watts used for one hour

NCAR National Center for Atmospheric Research, a US government agency involved in climate models

NCDC National Climactic Data Center

net-zero carbon a method of energy generation in which the carbon dioxide emitted in producing energy is balanced by carbon sinks that are specifically created to offset it. Carbon sinks (+), such as forests, and carbon emissions (-) are added together to reach zero.

NGO nongovernmental organization: An organized, highly financed group that tries to achieve social or political aims but supposedly is not controlled by a government

NOAA National Oceanic and Atmospheric Administration, or "National Weather Service", part of the US Dept. of Commerce

peer review evaluation of a scientist's work by qualified people working in the same scientific discipline

proxy data climate data that is derived from clues other than weather instruments, such as ice core samples, tree rings, etc.

radiative forcing the effect of greenhouse gases on the transfer of heat in the atmosphere

RED [European Union] Renewable Energy Directives

RSS remote sensing system (satellites)

scientism an exaggerated, unrealistic trust in the ability of natural science to solve all of our problems

solar wind a stream of protons originating on the sun that, when arriving at Earth, help deflect cosmic rays from outer space, thus preventing the formation of low clouds that block solar radiation and cool the climate

TSI total solar irradiance

TWh terawatt hour; one thousand gigawatts used for one hour

UNEP United Nations Environment Program

UNFCCC United Nations Framework Convention on Climate Change

USHCN United States Historical Climatology Network

WMO World Meteorological Organization

Appendix II

Some Talking Points

1. How is it that in the 1960s and 70s when the world's carbon dioxide emissions were rising rapidly, the earth was cooling?

2. During the Medieval Warm Period, 1000 years ago, Earth's average temperature was about two degrees higher than it now is. Humans weren't burning fossil fuels at that time. What caused that temperature increase?

3. What source of energy do you propose to replace the combustion of carbon-based "fossil" fuels, when experience has shown that wind and solar generated energy is undependable, devastating to the environment and three times as expensive?

4. How will solar and wind power help the US economy, when electricity produced by these methods is at least three times as expensive as carbon based and nuclear fuels?

5. How will solar and wind energy power all of the cars, trucks, farm tractors and heavy construction vehicles that are currently used in the US economy?

6. If the global warming hypothesis is valid, why have agencies such as NOAA and NASA found it necessary to change temperature records by lowering earlier temperatures and raising later temperatures?

7. For what length of time can batteries store the energy we'll need when the wind isn't blowing and the sun isn't shining? (ANSWER: about 4 hours)

8. Did you ever hear about Climategate (in 2009 and 2011)? (Chapter 2)

9. How did Al Gore, prime spokesman for climate change, increase his net worth from two million dollars in 2001 to over $300 million in 2022?

10. How can electric cars save on carbon dioxide emissions when being recharged with electricity made using coal or natural gas?

11. Where will the electricity needed to power electric vehicles come from when all the vehicles on the highway are electric?

12. Why are Germany, Spain and other European countries that were dedicated to "going green" starting up their coal-fired electric generators again?

13. Do climatologists use ground based temperature records or satellite temperature monitoring to determine the earth's average temperature?

14. Why do experts disagree about the causes and effects of climate change?

15. Did you know that recently – from 2005 to 2017 – a 12-year period, no major hurricane hit the coast of the United States? This was the longest pause in hurricane activity in recent history. So how can climate change alarmists say there are now more hurricanes than before?

16. If wind and solar energy are free as they say, why has the cost of electricity in the United Kingdom and Germany, which invested heavily in wind and solar, doubled?

17. How many hours worth of electricity can lithium-ion batteries store? (4 hours)

18. Why does John Kerry, the climate czar of the Biden Regime, fly all over the world in a private jet that produced 325 metric tons of carbon in the first year of Biden's presidency?

19. Why would Barack Obama, who told us the sea level was skyrocketing up, buy two oceanfront properties after leaving office?

20. What is the most powerful greenhouse gas? (ANSWER: water vapor)

Appendix III

Quotes By Top Scientists

1. "I have never witnessed a more disturbing corruption of the peer review process than events that led to this IPCC report." Frederick Seitz, IPCC scientist and former head of the US National Academy of Sciences, the American Physical Society and Rockefeller University.

2. James Lovelock, who proposed the Gaia hypothesis and once believed in AGW, stated, "I'm not sure the whole thing isn't crazy."

3. "None of the strategies that have been offered by the US government or by the EPA or by anybody else has the remotest chance of altering climate, if in fact climate is controlled by carbon dioxide." Dr. Robert Giegengack, University of Pennsylvania geologist

4. "Global warming is a nonproblem...I say this to Obama: Excuse me, Mr. President, but you're wrong. Dead wrong...Global warming really has become a new religion." Dr. Ivar Giaever, Nobel prize winner who had endorsed Obama

5. "It disturbs me hearing people like Al Gore say that he is worried about the future of his grandchildren. Complete madness." Dr. Richard Tol, UN IPCC Lead Author

6. "Global warming fears are the worst scientific scandal in history...When people come to know what the truth is, they will feel deceived by science and scientists." Dr. Kiminori Itoh, PhD, award-winning environmental physical chemist

7. "To call carbon dioxide a pollutant is really Orwellian. You are calling something a pollutant that we all produce. Where does that lead us eventually?" Dr. Will Happer, Princeton physicist

8. "We had both higher temperatures and an ice age at a time when CO_2 emissions were ten times higher than they are today." Dr. Patrick Moore, co-founder of Greenpeace

Made in United States
Troutdale, OR
05/09/2025

31218393R00071